Introduction to Soil Science

Introduction to Soil Science

Ron Schultz

SYRAWOOD
PUBLISHING HOUSE

New York

Published by Syrawood Publishing House,
750 Third Avenue, 9th Floor,
New York, NY 10017, USA
www.syrawoodpublishinghouse.com

Introduction to Soil Science
Ron Schultz

International Standard Book Number: 978-1-64740-007-1 (Hardback)

This book contains information obtained from authentic and highly regarded sources. All chapters are published with permission under the Creative Commons Attribution Share Alike License or equivalent. A wide variety of references are listed. Permissions and sources are indicated; for detailed attributions, please refer to the permissions page. Reasonable efforts have been made to publish reliable data and information, but the authors, editors and publisher cannot assume any responsibility for the validity of all materials or the consequences of their use.

Trademark Notice: Registered trademark of products or corporate names are used only for explanation and identification without intent to infringe.

Cataloging-in-Publication Data

Introduction to soil science / Ron Schultz.
    p. cm.
Includes bibliographical references and index.
ISBN 978-1-64740-007-1
1. Soil science. 2. Agriculture. 3. Soils. I. Schultz, Ron.
S591 .I58 2020
631.4--dc23

TABLE OF CONTENTS

PREFACE

Soil science is the study of soil, including its formulation, classification, and mapping. It examines the physical, biological, chemical and fertility properties of different types of soils available on the earth's surface. Soil science studies such properties concerning the use and management of soils. The two main branches of soil science are pedology and edaphology. Pedology deals with the formation, morphology, chemistry, and classification of soil. Edaphology is concerned with the interaction of soil with living things, particularly plants. Some of the areas of study under this discipline include soil genesis, soil morphology, soil microbiology, soil mechanics and agricultural soil science. This textbook explores all the important aspects of soil science in the present day scenario. It elucidates new techniques and their applications in a multidisciplinary approach. The coherent flow of topics, student-friendly language and extensive use of examples make this book an invaluable source of knowledge.

To facilitate a deeper understanding of the contents of this book a short introduction of every chapter is written below:

Chapter 1- Soil is a material composed of minerals, gases, organic matter, liquids and organisms. Pedosphere is the earth's body of soil which plays a major role in growing plants, water storage supply and purification, habitat for organisms and as a modifier of earth's atmosphere. Clay soil, sandy soil, peat soil, loam soil, chalky soil, etc. are the various types of soil. This is an introductory chapter which will introduce briefly the classification of soil.

Chapter 2- The process of soil formation regulated by the effects of place, environment, and history is known as pedogenesis. There are various factors that affect soil formation such as organisms, climate and parental materials, relief and time. In order to completely understand the nature of soil, it is necessary to understand the processes related to it. The following chapter elucidates the varied processes and mechanisms associated with this area of study.

Chapter 3- Soil physics is a domain that deals with the study of physical properties and processes of soil. It also deals with the phases of physical soil components and their dynamics. The study of soil texture, soil structure, organic matter in soil and soil composition fall under this domain. This chapter closely examines these key concepts of soil physics to provide an extensive understanding of the subject.

Chapter 4- The study of the chemical characteristics of soil is known as soil chemistry. It deals with the study of soil reaction such as oxidation reaction, soil ph, soil salinity, soil sodicity, etc. The factors that affect soil chemistry are mineral composition, environmental factors and organic matter. This chapter discusses in detail the diverse aspects related to soil chemistry.

Chapter 5- The study of microbial and faunal activity and ecology in soil is termed as soil biology. It is primarily concerned with soil respiration, soil biological crust, soil regulators and soil system. Soil ecology is a domain that studies the interactions between soil biology and abiotic and biotic facets of the soil environment. The topics elaborated in this chapter will help in gaining a better perspective about the key concepts of soil biology and soil ecology.

Finally, I would like to thank the entire team involved in the inception of this book for their valuable time and contribution. This book would not have been possible without their efforts. I would also like to thank my friends and family for their constant support.

Ron Schultz

Chapter 1

Soil and its Classification

Soil is a material composed of minerals, gases, organic matter, liquids and organisms. Pedosphere is the earth's body of soil which plays a major role in growing plants, water storage supply and purification, habitat for organisms and as a modifier of earth's atmosphere. Clay soil, sandy soil, peat soil, loam soil, chalky soil, etc. are the various types of soil. This is an introductory chapter which will introduce briefly the classification of soil.

Soil is the biologically active, porous medium that has developed in the uppermost layer of Earth's crust. Soil is one of the principal substrata of life on Earth, serving as a reservoir of water and nutrients, as a medium for the filtration and breakdown of injurious wastes, and as a participant in the cycling of carbon and other elements through the global ecosystem. It has evolved through weathering processes driven by biological, climatic, geologic, and topographic influences.

Since the rise of agriculture and forestry in the 8th millennium BCE, there has also arisen by necessity a practical awareness of soils and their management. In the 18th and 19th centuries the Industrial Revolution brought increasing pressure on soil to produce raw materials demanded by commerce, while the development of quantitative science offered new opportunities for improved soil management. The study of soil as a separate scientific discipline began about the same time with systematic investigations of substances that enhance plant growth. This initial inquiry has expanded to an understanding of soils as complex, dynamic, biogeochemical systems that are vital to the life cycles of terrestrial vegetation and soil-inhabiting organisms—and by extension to the human race as well.

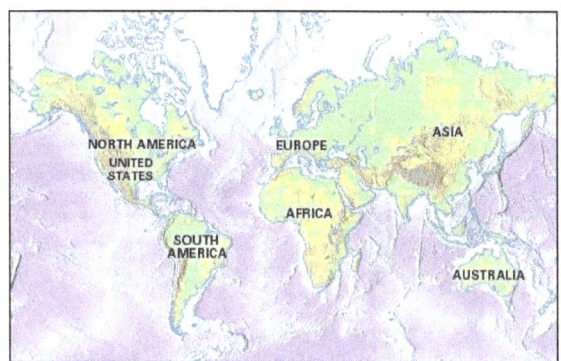
Interactive soil map of the world.

The Soil Profile

Soil Horizons

Soils differ widely in their properties because of geologic and climatic variation over distance and time. Even a simple property, such as the soil thickness, can range from a few centimetres to many metres, depending on the intensity and duration of weathering, episodes of soil deposition and erosion, and the patterns of landscape evolution. Nevertheless, in spite of this variability, soils have a unique structural characteristic that distinguishes them from mere earth materials and serves as a basis for their classification: a vertical sequence of layers produced by the combined actions of percolating waters and living organisms.

Podzol soil profile from Ireland, showing a bleached layer from which humus and metal oxides have been leached and subsequently deposited in the typically reddish horizon below.

These layers are called horizons, and the full vertical sequence of horizons constitutes the soil profile. Soil horizons are defined by features that reflect soil-forming processes. For instance, the uppermost soil layer (not including surface litter) is termed the A horizon. This is a weathered layer that contains an accumulation of humus (decomposed, dark-coloured, carbon-rich matter) and microbial biomass that is mixed with small-grained minerals to form aggregate structures.

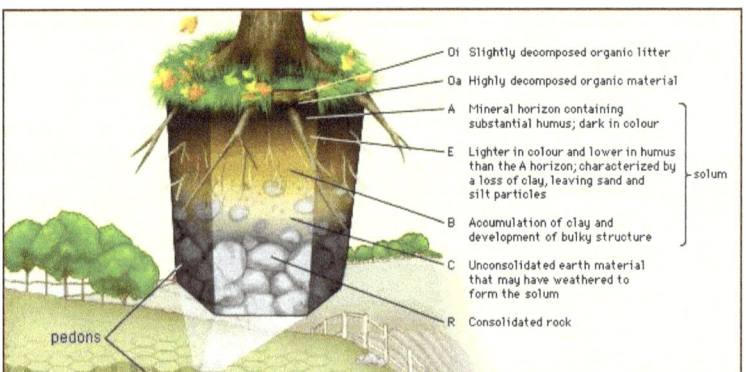

The soil profile, showing the major layers from the O horizon (organic material) to the R horizon (consolidated rock). A pedon is the smallest unit of land surface that can be used to study the characteristic soil profile of a landscape.

Below A lies the B horizon. In mature soils this layer is characterized by an accumulation of clay (small particles less than 0.002 mm [0.00008 inch] in diameter) that has either been deposited out of percolating waters or precipitated by chemical processes involving dissolved products of weathering. Clay endows B horizons with an array of diverse structural features (blocks, columns, and prisms) formed from small clay particles that can be linked together in various configurations as the horizon evolves.

Below the A and B horizons is the C horizon, a zone of little or no humus accumulation or soil structure development. The C horizon often is composed of unconsolidated parent material from which the A and B horizons have formed. It lacks the characteristic features of the A and B horizons and may be either relatively unweathered or deeply weathered. At some depth below the A, B, and C horizons lies consolidated rock, which makes up the R horizon.

These simple letter designations are supplemented in two ways. First, two additional horizons are defined. Litter and decomposed organic matter (for example, plant and animal remains) that typically lie exposed on the land surface above the A horizon are given the designation O horizon, whereas the layer immediately below an A horizon that has been extensively leached (that is, slowly washed of certain contents by the action of percolating water) is given the separate designation E horizon, or zone of eluviation. The development of E horizons is favoured by high rainfall and sandy parent material, two factors that help to ensure extensive water percolation. The solid particles lost through leaching are deposited in the B horizon, which then can be regarded as a zone of illuviation.

Soil horizon letter designations	
Base symbols for surface horizons	
O	organic horizon containing litter and decomposed organic matter
A	mineral horizon darkened by humus accumulation
Base symbols for subsurface horizons	
E	mineral horizon lighter in colour than an A or O horizon and depleted in clay minerals
AB or EB	transitional horizon more like A or E than B
BA or BE	transitional horizon more like B than A or E
B	accumulated clay and humus below the A or E horizon
BC or CB	transitional horizon from B to C
C	unconsolidated earth material below the A or B horizon
R	consolidated rock
Suffixes added for special features of horizons	
a	highly decomposed organic matter
b	buried horizon
c	concretions or hard nodules (iron, aluminum, manganese, or titanium)
e	organic matter of intermediate decomposition
f	frozen soil

g	gray colour with strong mottling and poor drainage
h	accumulation of organic matter
i	slightly decomposed organic matter
k	accumulation of carbonate
m	cementation or induration
n	accumulation of sodium
o	accumulation of oxides of iron and aluminum
p	plowing or other anthropogenic disturbance
q	accumulation of silica
r	weathered or soft bedrock
s	accumulation of metal oxides and organic matter
t	accumulation of clay
v	plinthite (hard iron-enriched subsoil material)
w	development of colour or structure
x	fragipan character (high-density, brittle)
y	accumulation of gypsum
z	accumulation of salts

The combined A, E, B horizon sequence is called the solum (Latin: "floor"). The solum is the true seat of soil-forming processes and is the principal habitat for soil organisms. (Transitional layers, having intermediate properties, are designated with the two letters of the adjacent horizons.)

The second enhancement to soil horizon nomenclature is the use of lowercase suffixes to designate special features that are important to soil development. The most common of these suffixes are applied to B horizons: g to denote mottling caused by waterlogging, h to denote the illuvial accumulation of humus, k to denote carbonate mineral precipitates, o to denote residual metal oxides, s to denote the illuvial accumulation of metal oxides and humus, and t to denote the accumulation of clay.

Pedons and Polypedons

Soils are natural elements of weathered landscapes whose properties may vary spatially. For scientific study, however, it is useful to think of soils as unions of modules known as pedons. A pedon is the smallest element of landscape that can be called soil. Its depth limit is the somewhat arbitrary boundary between soil and "not soil" (e.g., bedrock). Its lateral dimensions must be large enough to permit a study of any horizons present—in general, an area from 1 to 10 square metres (10 to 100 square feet), taking into account that a horizon may be variable in thickness or even discontinuous. Wherever horizons are cyclic and recur at intervals of 2 to 7 metres (7 to 23 feet), the pedon includes one-half the cycle. Thus, each pedon includes the range of horizon variability that occurs within small areas. Wherever the cycle is less than 2 metres, or wherever all horizons are continuous and of uniform thickness, the pedon has an area of 1 square metre.

Soils are encountered on the landscape as groups of similar pedons, called polypedons, that contain sufficient area to qualify as a taxonomic unit. Polypedons are bounded from below by "not soil" and laterally by pedons of dissimilar characteristics.

Soil Behaviour

Physical Characteristics

- Grain Size and Porosity

The grain size of soil particles and the aggregate structures they form affect the ability of a soil to transport and retain water, air, and nutrients. Grain size is classified as clay if the particle diameter is less than 0.002 mm (0.0008 inch), as silt if it is between 0.002 mm (0.0008 inch) and 0.05 mm (0.002 inch), or as sand if it is between 0.05 mm (0.002 inch) and 2 mm (0.08 inch). Soil texture refers to the relative proportions of sand, silt, and clay particle sizes, irrespective of chemical or mineralogical composition. Sandy soils are called coarse-textured, and clay-rich soils are called fine-textured. Loam is a textural class representing about one-fifth clay, with sand and silt sharing the remainder equally.

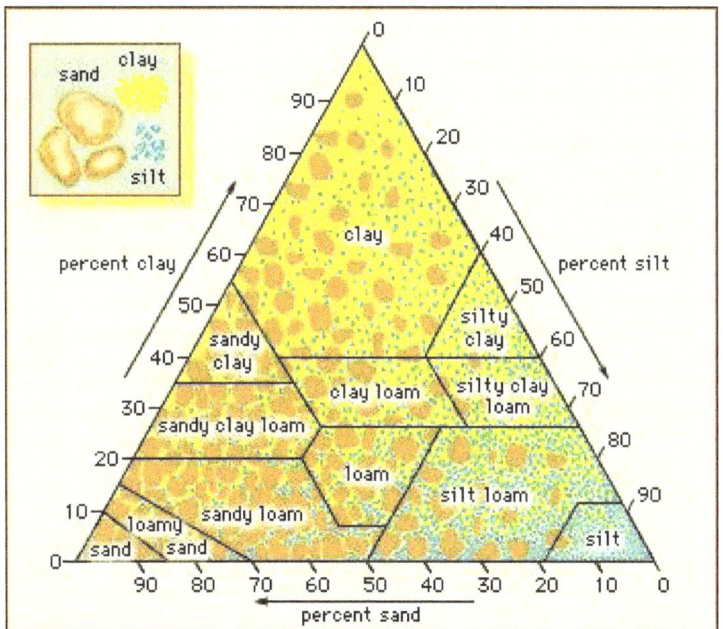

Soil texture as a function of the proportion of sand, silt, and clay particle sizes.

Pore radii (space between soil particles) can range from millimetre-scale between sand grains to micrometre-scale between clay grains. Soil particles falling into the three principal size categories may have various mineralogical or chemical compositions, although sand particles often are composed of quartz and feldspars, silt particles often are micaceous, and clay particles often contain layer-type aluminosilicates (the so-called clay minerals). Organic matter and amorphous mineral matter also are important constituents of soil clay particles.

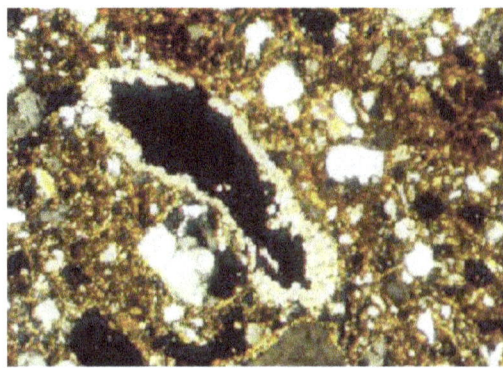

Microscopic view of an Inceptisol, showing small crystallites of carbonate minerals (around the central black void), quartz sand grains (white), and iron oxides and organic matter (dark brown).

Porosity reflects the capacity of soil to hold air and water, and permeability describes the ease of transport of fluids and their dissolved components. The porosity of a soil horizon increases as its texture becomes finer, whereas the permeability decreases as the average pore size becomes smaller. Small pores not only restrict the passage of matter, but they also bring it into close proximity with chemical binding sites on the particle surface that can slow its movement. Clay and humus affect both soil porosity and permeability by binding soil grains together into aggregates, thereby creating a network of larger pores (macropores) that facilitate the movement of water. Plant roots open pores between soil aggregates, and cycles of wetting and drying create channels that allow water to pass easily. (However, this structure collapses under waterlogging conditions.) The stability of aggregates increases with humus content, especially humus that originates from grass vegetation. For soils that are not disturbed significantly by human activities, however, the pore space and the varieties of macropores are more important determinants of porosity than the soil texture. As a general rule, average pore size decreases from certain agricultural practices and other human uses of soil.

- Water Runoff

Aggregates of soil particles whose formation has not been influenced by human intervention are called peds. The peds in the surface horizons of soils develop into clods under the effects of cultivation and the traffic of urbanization. Soils whose A horizon is dense and unstructured increase the fraction of precipitation that will become surface runoff and have a high potential for erosion and flooding. These soils include not only those whose peds have been degraded but also coarse-textured soils with low porosity, particularly those of arid regions.

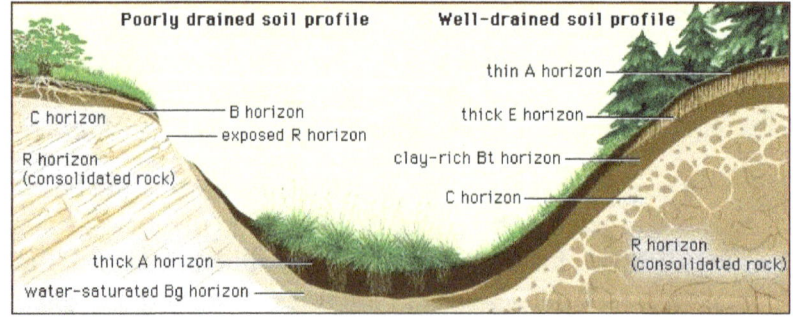

Soil profiles on hillslopes

The thickness and composition of soil horizons vary with position on a hillslope and with water drainage. For example, on the upper slopes of poorly drained profiles, underlying rock may be exposed by surface erosion, and nutrient-rich soils (A horizon) may accumulate at the toeslope. On the other hand, in well-drained profiles under forest cover, the leached layers (E horizon) may be relatively thick and surface erosion minimal.

A well-developed clay horizon (Bt) presents a deep-lying obstacle to the downward percolation of water. Subsurface runoff cannot easily penetrate the clay layer and flows laterally along the horizon as it moves toward the stream system. This type of runoff is slower than its erosive counterpart over the land surface and leads to water saturation of the upper part of the soil profile and the possibility of gravity-induced mass movement on hillslopes (e.g., landslides). It is also responsible for the translocation (migration) of dissolved products of chemical weathering down a hillslope sequence of related soil profiles (a toposequence). Subsurface water flow is also influenced by macropores, which, are created through plant root growth and decay, animal burrowing activities, soil shrinkage while drying, or fracturing. In general, subsurface runoff processes are characteristic of soils in humid regions, whereas surface runoff is characteristic of arid regions and, of course, any landscape altered significantly by cultivation or urbanization.

Chemical Characteristics

Mineral Content

The bulk of soil consists of mineral particles that are composed of arrays of silicate ions (SiO_4^{4-}) combined with various positively charged metal ions. It is the number and type of the metal ions present that determine the particular mineral. The most common mineral found in Earth's crust is feldspar, an aluminosilicate that contains sodium, potassium, or calcium (sometimes called bases) in addition to aluminum ions. Weathering breaks up crystals of feldspars and other silicate minerals and releases chemical compounds such as bases, silica, and oxides of iron and aluminum (Fe_2O_3 and alumina [Al_2O_3]). After the bases are removed by leaching, the remaining silica and alumina combine to form crystalline clays.

The kind of crystalline clay produced depends on leaching intensity. Prolonged leaching leaves little silica to combine with alumina and results in what are known as 1:1 clays, consisting of alternating silica and alumina sheets; less extensive leaching leads to the formation of 2:1 clays, consisting of one alumina sheet sandwiched between two silica sheets. In neither case is the result solely one of the two types, though 1:1 clay is predominant in the tropics after prolonged leaching and 2:1 clay more abundant when leaching is less extensive in more temperate climates.

The solid soil particles are chemically reactive because of the presence of electrically charged sites on their surfaces. If a reactive site binds a dissolved ion or molecule to form a stable unit, a "surface complex" is said to exist. The formation reaction itself is called surface complexation. Surface complexation is an example of adsorption, a chemical process in which matter accumulates on a solid particle surface. Ions such as Ca^{2+} (calcium), Mg^{2+} (magnesium), Na^+ (sodium), and NO_3^- (nitrate) do not tend to adsorb strongly, making these important plant nutrients susceptible to easy replacement. Once ejected from their surface sites, these ions may be leached downward by percolating water to become removed from the biogeochemical cycles occurring in the upper part of the soil profile.

Freshwater leaching of soils brings hydrogen ions (H^+) that increase mineral solubility, releasing Al^{3+}(aluminum), a toxic ion that can displace nutrients such as Ca^{2+}. The gradual loss of nutrients and the accumulation of adsorbed H^+ and Al^{3+} characterize the buildup of soil acidity, with its harmful effects on organisms. Soils display their acidity by a decrease in content of acid-soluble minerals (for example, feldspars or clay minerals) and an increase in insoluble minerals (iron and aluminum oxides). Soils weathered by freshwater leaching evolve from clay particles with a prevalence of metal ion-binding sites to highly weathered metal oxides that do not have sites that bind readily with metal ions.

Ferralsol soil profile from Brazil, showing a deep red subsurface horizon
resulting from accumulations of iron and aluminum oxides.

Organic Content

The second major component of soils is organic matter produced by organisms. The total organic matter in soil, except for materials identifiable as undecomposed or partially decomposed biomass, is called humus. This solid, dark-coloured component of soil plays a significant role in the control of soil acidity, in the cycling of nutrients, and in the detoxification of hazardous compounds. Humus consists of biological molecules such as proteins and carbohydrates as well as the humic substances (polymeric compounds produced through microbial action that differ from metabolically active compounds).

Mollisol soil profile, showing a typically dark surface horizon rich in humus.

The processes by which humus forms are not fully understood, but there is agreement that four stages of development occur in the transformation of soil biomass to humus:

1. Decomposition of biomass into simple organic compounds,

2. Metabolization of the simple compounds by microbes,

3. Cycling of carbon, hydrogen, nitrogen, and oxygen between soil organic matter and the microbial biomass, and

4. Microbe-mediated polymerization of the cycled organic compounds.

The investigation of molecular structure in humic substances is a difficult area of current research. Although it is not possible to describe the molecular configuration of humic substances in any but the most general terms, these molecules contain hydrogen ions that dissociate in fresh water to form molecules that bear a net negative charge. These negatively charged sites can interact with toxic metal ions and effectively remove them from further interaction with the environment.

Much of the molecular framework of soil organic matter, however, is not electrically charged. The uncharged portions of humic substances can react with synthetic organic compounds such as pesticides, fertilizers, solid and liquid waste materials, and their degradation products. Humus, either as a separate solid phase or as a coating on mineral surfaces, can immobilize these compounds and, in some instances, detoxify them significantly.

Biological Phenomena

Fertile soils are biological environments teeming with life on all size scales, from microfauna (with body widths less than 0.1 mm [0.004 inch]) to mesofauna (up to 2 mm [0.08 inch] wide) and macrofauna (up to 20 mm [0.8 inch] wide). The most numerous soil organisms are the unicellular microfauna: 1 kilogram (2.2 pounds) of soil may contain 500 billion bacteria, 10 billion actinomycetes (filamentous bacteria, some of which produce antibiotics), and nearly 1 billion fungi. The multicellular animal population can approach 500 million in a kilogram of soil, with microscopic nematodes (roundworms) the most abundant. Mites and springtails, which are categorized as mesofauna, are the next most prevalent. Earthworms, millipedes, centipedes, and insects make up most of the rest of the larger soil animal species. Plant roots also make a significant contribution to the biomass—the combined root length from a single plant can exceed 600 km (373 miles) in the top metre of a soil profile.

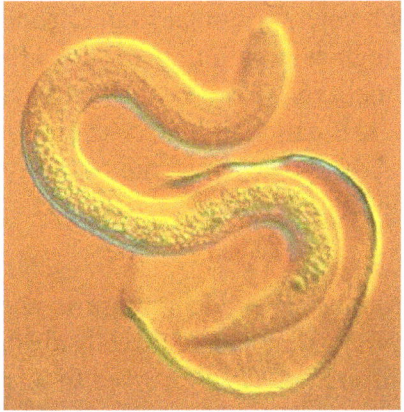

Roundworm (nematode) hatching. Nematodes feed on bacteria, fungi, and small animal forms in the soil

The soil flora and fauna play an important role in soil development. Microbiological activity in the rooting zone of soils is important to soil acidity and to the cycling of nutrients. Aerobic and anaerobic (oxygen-depleted) microniches support microbes that determine the rate of the production of carbon dioxide(CO_2) from organic matter or of nitrate (NO_3^-) from molecular nitrogen (N_2).

The carbon and nitrogen cycles are two important microbe-mediated cycles. In this topic, however, it is worth pointing out how they illustrate the complex, integrated nature of a soil's physical, chemical, and biological behaviour: soil peds and pore spaces provide microniches for the action of carbon- and nitrogen-cycling organisms, soil humus provides the nutrient reservoirs, and soil biomass provides the chemical pathways for cycling. The carbon in dead biomass is converted to CO_2 by aerobic microorganisms and to organic acids or alcohols by anaerobic microorganisms. Under highly anaerobic conditions, methane (CH_4) is produced by bacteria. The CO_2 produced can be used by photosynthetic microorganisms or by higher plants to create new biomass and thus initiate the carbon cycle again.

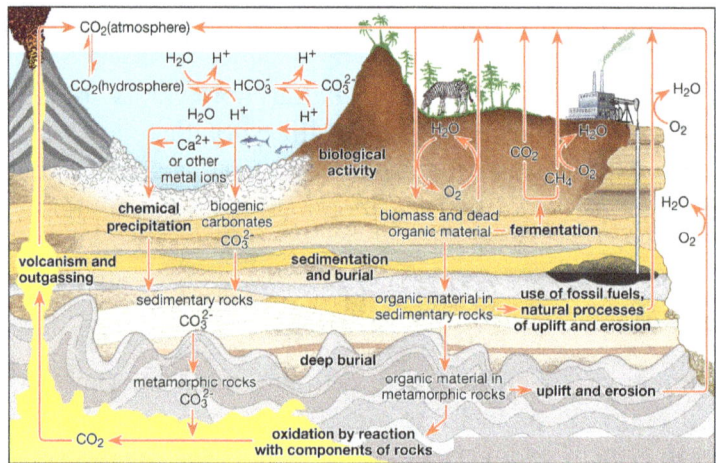

The carbon cycle

Carbon is transported in various forms through the atmosphere, the hydrosphere, and geologic formations. One of the primary pathways for the exchange of carbon dioxide (CO_2) takes place between the atmosphere and the oceans; there a fraction of the CO_2 combines with water, forming carbonic acid (H_2CO_3) that subsequently loses hydrogen ions (H^+) to form bicarbonate (HCO_3^-) and carbonate (CO_3^{2-}) ions. Mollusk shells or mineral precipitates that form by the reaction of calcium or other metal ions with carbonate may become buried in geologic strata and eventually release CO_2 through volcanic outgassing. Carbon dioxide also exchanges through photosynthesis in plants and through respiration in animals. Dead and decaying organic matter may ferment and release CO_2 or methane (CH_4) or may be incorporated into sedimentary rock, where it is converted to fossil fuels. Burning of hydrocarbon fuels returns CO_2 and water (H_2O) to the atmosphere. The biological and anthropogenic pathways are much faster than the geochemical pathways and, consequently, have a greater impact on the composition and temperature of the atmosphere.

The nitrogen (N) bound into proteins in dead biomass is consumed by microorganisms and converted into ammonium ions (NH_4^+) that can be directly absorbed by plant roots (for example, lowland rice). The ammonium ions are usually converted to nitrite ions (NO_2^-) by *Nitrosomonas* bacteria, followed by a second conversion to nitrate (NO_3^-) by Nitrobacter bacteria. This very mobile form of nitrogen is that most commonly absorbed by plant roots, as well as by microorganisms in soil. To close the nitrogen cycle, nitrogen gas in the atmosphere is converted to biomass nitrogen by Rhizobium bacteria living in the root tissues of legumes (e.g., alfalfa, peas, and beans) and leguminous trees (such as alder) and by cyanobacteria and Azotobacter bacteria.

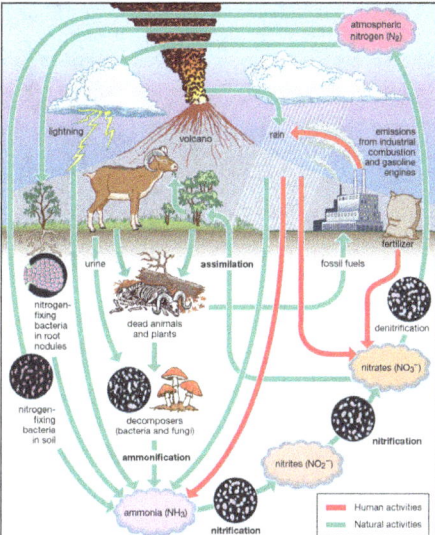

The nitrogen cycle

Soil Erosion

Soil profiles are continually disrupted by the actions of flowing water, wind, or ice and by the force of gravity. These erosive processes remove soil particles from A horizons and expose subsurface horizons to weathering, resulting in the loss of humus, plant nutrients, and beneficial soil organisms. Not only are these losses of paramount importance to agriculture and forestry, but the removal, transport, and subsequent deposition of soil can have significant economic consequences by damaging buildings, bridges, culverts, and other structures.

Erosive Processes

Water-induced erosion can take various forms depending on climate and topography. The force of rainfall striking a land surface unimpeded by vegetation or man-made structures is sufficient to raise 15 cm (6 inches) of material from an A horizon nearly 1 metre (39 inches) into the air. The impact of raindrops breaks the bonds holding soil aggregates together and catapults the particles into the flowing water from surface runoff. Wholesale removal of soil particles by the sheet flow of water (sheet erosion) or by flow in small channels (rill erosion) accounts for most of the water-induced soil loss from exposed land surfaces. More spectacular but less prevalent types of erosion are gully erosion, in which water concentrates in channels too deep to smooth over by tilling, and streambank erosion, in which the saturated sides of running streams tumble into the moving water below. The same forces at work in streambank erosion are seen in soils on hillslopes that become thoroughly saturated with water. Gravity, able to overcome the cohesive forces that hold soil particles together, can cause the entire soil profile to move downslope—a phenomenon called mass movement. This movement may be either slow (soil creep), rapid (debris flow or mudflow), or sometimes catastrophic (landslide).

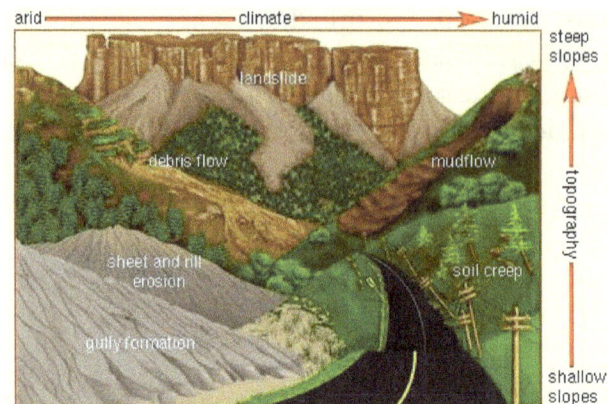

Effect of topography and climate on water-induced soil erosion

On shallow slopes the predominant forms of erosion in arid climates are gully formation or sheet and rill erosion, whereas soil creep is seen in more humid climates. As the slope of the terrain becomes steeper, mudflows, debris flows, and landslides become the primary modes of erosion.

The mechanisms involved in wind erosion depend on soil texture and the size of soil particles. Dry soil particles of silt or clay size can be transported over great distance by wind. Larger particles that are the size of fine sand, 0.05 mm (0.002 inch) to 0.5 mm (0.02 inch) in diameter, can be vaulted as high as 25 cm (10 inches) into the air, then drop to the ground after a short flight, only to rebound under the continual driving force of the wind. Coarser sand particles are not lifted, but they can tumble along the land surface. The major cause of wind erosion is the jumping motion of the smaller soil particles, a process called saltation. The texture of the windblown surfaces of these soils becomes coarser, making them less chemically reactive and less able to retain plant nutrients or trap pollutants. In arid regions, wind erosion often produces a gravelly land surface known as desert pavement.

Rates of Soil Erosion

Soil erosion and deposition are natural geomorphic processes that give shape to landforms and provide new parent material for the development of soil profiles. These processes become soil conservation issues when the rate of erosion greatly exceeds the rate expected in the absence of human land use—a situation referred to as accelerated erosion. Rates of normal soil erosion have been estimated from measurements of sediment transport and accumulation, mass movement on hillslopes, and radioactive carbon dating of landforms. They range from less than 0.02 to more than 10 metric tons per hectare (0.01 to 4.5 tons per acre) of soil lost annually. In comparison the rates of natural soil formation range from 0.2 to 9 metric tons per hectare per year. The average annual rate of normal soil erosion is nearly 1 metric ton per hectare (0.45 ton per acre), while that of natural soil formation is nearly 0.7 metric ton per hectare (0.3 ton per acre). Broad variation is the rule, but rates of soil loss exceeding 10 metric tons per hectare annually signal accelerated erosion. It is important to note that this accelerated soil loss is equivalent to less than 1 mm (0.04 inch) of soil depth, making erosion damage very difficult to observe over short time spans.

When climate and topography are fixed and soil cover is varied, the rate of soil loss by water erosion has a predictable and dramatic dependence on vegetation. Irrespective of location, erosion losses are usually very small from forestland or permanent pastureland, moderate to high from land planted with grain crops, and very high from clean-tilled orchards, vineyards, and land planted with row crops, as shown in the figure.

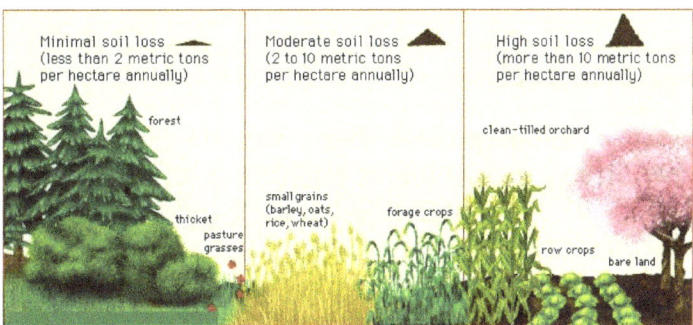

Soil loss versus vegetative coverThe amount of topsoil lost by water erosion depends on the amount and type of vegetation. Forests and grasslands lose significantly less soil by erosion than do highly cultivated lands.

Resistance to Erosion

The ability of soils to resist water and wind erosion depends on their texture and topographic characteristics. Clay-rich soils resist erosion well because of strong cohesive forces between particles and the gluelike characteristics of humus. Both loam and sandy soils are moderately resistant to erosion—the former because they have sufficient clay content to hold the particles together, the latter because their high permeability limits the amount of surface runoff that can wash soil particles away, while their larger particle size makes them too heavy to be easily entrained (transported) in flowing water. Silty soils, on the other hand, exhibit the least resistance to erosion because their permeability is low (resulting in more surface runoff), and their particle size is neither small enough to promote cohesion nor large enough to prevent entrainment. Soils on steep, long slopes are much more susceptible to erosion than those on shallow, short slopes because the steeper slopes accelerate the flow of surface runoff.

The development of soil conservation strategies requires knowledge of actual and acceptable rates of soil erosion. A practical measure of soil resistance to erosion used by pedologists in the United States is the soil loss tolerance (T-value, or T-factor). This quantity is defined as the maximum annual rate of soil loss by erosion that will permit high soil productivity for an indefinite period of time. Operationally, the concept is interpreted as the maximum annual loss from the A horizon that does not reduce the thickness of the rooting zone significantly over millennia.

Guidelines have been developed by the U.S. Natural Resource Conservation Service to assist field estimations of the T-value based on texture, topography, and depth to bedrock or to a root-impeding layer (hardpan) in a soil profile. Deep, coarse-textured soils are assigned a T-value of 11.2 metric tons per hectare (5 tons per acre), fine-textured soils have a T-value of 9 metric tons per hectare (4 tons per acre), and shallow soils or those with an impeding layer are assigned T-values in the range of 2.2–6.7 metric tons per hectare (1–3 tons per acre), depending on texture. Unfavourable slope characteristics are used to modify these values downward as experience may warrant.

Soils in Ecosystems

An ecosystem is a collection of organisms and the local environment with which they interact. For the soil scientist studying microbiological processes, ecosystem boundaries may enclose a single soil horizon or a soil profile. When nutrient cycling or the effects of management practices on soils are being considered, the ecosystem may be as large as an entire plant community and soil polypedon system.

Carbon and Nitrogen Cycles

Soils are dynamic, open habitats that provide plants with physical support, water, nutrients, and air for growth. Soils also sustain an enormous population of microorganisms such as bacteria and fungi that recycle chemical elements, notably carbon and nitrogen, as well as elements that are toxic. The carbonand nitrogen cycles are important natural processes that involve the uptake of nutrients from soil, the return of organic matter to the soil by tissue aging and death, the decomposition of organic matter by soil microbes (during which nutrients or toxins may be cycled within the microbial community), and the release of nutrients into soil for uptake once again. These cycles are closely linked to the hydrologic cycle, since water functions as the primary medium for chemical transport.

Nitrogen (N), one of the major nutrients, originates in the atmosphere. It is transformed and transported through the ecosystem by the water cycle and biological processes. This nutrient enters the biosphereprimarily as wet deposition to the soil surface (throughfall), where plants, microbial decomposers, or nitrifiers (microbes that convert ammonium [NH_4^+] to nitrate [NO_3^-]) compete for it. This competition plays a major role in determining the extent to which incoming nitrogen will be retained within an ecosystem.

Carbon (C) also enters the ecosystem from the atmosphere—in the form of carbon dioxide (CO_2)—and is taken up by plants and converted into biomass. Organic matter in the soil in the form of humus and other biomass contains about three times as much carbon as does land vegetation. Soils of arid and semiarid regions also store carbon in inorganic chemical forms, primarily as calcium carbonate ($CaCO_3$). These pools of carbon are important components of the global carbon cycle because of their location near the land surface, where they are subject to erosion and decomposition. Each year, soils release 4–5 percent of their carbon to the atmosphere by the transformation of organic matter into CO_2 gas, a process termed soil respiration. This amount of CO_2 is more than 10 times larger than that currently produced from the burning of fossil fuels (coal and petroleum), but it is returned to the soil as organic matter by the production of biomass.

A large portion of the soil carbon pool is susceptible to loss as a result of human activities. Land-use changes associated with agriculture can disrupt the natural balance between the production of carbon-containing biomass and the release of carbon by soil respiration. One estimate suggests that this imbalance alone results in an annual net release of CO_2 to the atmosphere from agricultural soils equal to about 20 percent of the current annual release of CO_2 from the burning of fossil fuels. Agricultural practices in temperate zones, for example, can result in a decline of soil organic matter that ranges from 20 to 40 percent of the original content after about 50 years of cultivation. Although a portion of this loss can be attributed to soil erosion, the majority is from an increased flux of carbon to the atmosphere as CO_2. The draining of peatlands may cause similarly large losses in soil carbon storage.

Soils and Global Change

Soils and climate have always been closely related. The predicted temperature increases due to global warming and the consequent change in rainfall patterns are expected to have a substantial impact on both soils and demographics. This anticipated climatic change is thought to be driven by the greenhouse effect—an increase in levels of certain trace gases in the atmosphere such as carbon dioxide (CO_2), methane (CH_4), and nitrous oxide (N_2O). The conversion of land to agriculture, especially in the humid tropics, is an important contribution to greenhouse gas emissions. Some computer models predict that CH_4 and N_2O emissions will also be very important in future global change. About 70 percent of the CH_4 and 90 percent of the N_2O in the atmosphere are derived from soil processes. But soils can also function as repositories for these gases, and it is important to appreciate the complexity of the source-repository relationship. For example, the application of nitrogen-containing fertilizers reduces the ability of the soil to process CH_4. Even the amount of nitrogen introduced into soil from acid rain on forests is sufficient to produce this effect. However, the extent of net emissions of CH_4 and N_2O and the microbial trade-off between the two gases are undetermined at the global scale.

Perhaps the most notable and pervasive role of soils in global change phenomena is the regulation of the CO_2 budget. Carbon that is stored in terrestrial plants mainly through photosynthesis is called net primary production or NPP and is the dominant source of food, fuel, fibre, and feed for the entire population of Earth. Approximately 55 billion metric tons (61 billion tons) of carbon are stored in this way each year worldwide, most of it in forests. About 800 million hectares (20 billion acres) of forestland have been lost since the dawn of civilization; this translates to about 6 billion metric tons of carbon per year less NPP than before land was cleared for agriculture and commerce. This estimated decrease in carbon storage can be compared to the 5–6 billion metric tons of carbon currently released per year by fossil fuelburning. One is left with the sobering conclusion that reforestation of the entire planet to primordiallevels would have only a temporary counterbalancing effect on carbon release to the atmosphere from human consumption of natural resources.

Carbon in terrestrial biomass that is not used directly becomes carbon in litter (about 25 billion metric tons of carbon annually) and is eventually incorporated into soil humus. Soil respiration currently releases an average of 68 billion metric tons of this carbon back into the atmosphere. The natural cycling of carbon is directly and indirectly affected by land-use changes through deforestation, reforestation, woodproducts decomposition, and abandonment of agricultural land. The current estimate of carbon loss from all these changes averages about 1.7 billion metric tons per year worldwide, or about one-third the current loss from fossil fuel burning. This figure could as much as double in the first half of the 21st century if the rate of deforestation is not controlled. Reforestation, on the other hand, could actually reduce the current carbon loss by up to 10 percent without exorbitant demands on management practices.

Basic Soil Components

A soil is simply a porous medium consisting of minerals, water, gases, organic matter, and microorganisms. The traditional definition is: Soil is a dynamic natural body having properties derived from the combined effects of climate and biotic activities, as modified by topography, acting on parent materials over time.

There are five basic components of soil that, when present in the proper amounts, are the backbone of all terrestrial plant ecosystems.

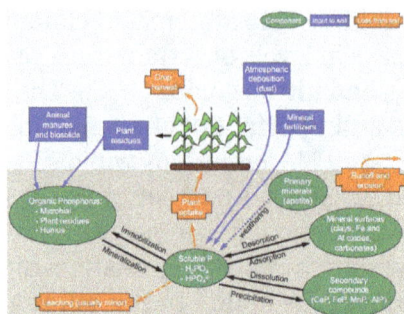

Soil is composed of a matrix of minerals, organic matter, air, and water. Each component is important for supporting plant growth, microbial communities, and chemical decomposition

Mineral

The largest component of soil is the mineral portion, which makes up approximately 45% to 49% of the volume. Soil minerals are derived from two principal mineral types. Primary minerals, such as those found in sand and silt, are those soil materials that are similar to the parent material from which they formed. They are often round or irregular in shape. Secondary minerals, on the other hand, result from the weathering of the primary minerals, which releases important ions and forms more stable mineral forms such as silicate clay. Clays have a large surface area, which is important for soil chemistry and water-holding capacity. Additionally, negative and neutral charges found around soil minerals influences the soil's ability to retain important nutrients, such as cations, contributing to a soils cation exchange capacity (CEC).

The texture of a soil is based on the percentage of sand, silt, and clay found in that soil. The identification of sand, silt, and clay are made based on size. The following is used in the United States:

- Sand 0.05 – 2.00 mm in diameter.

- Silt 0.002 – 0.05 mm in diameter.

- Clay < 0.002 mm in diameter.

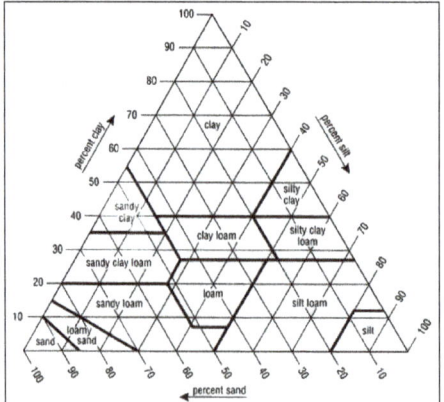

The U.S. Department of Agriculture Soil Texture Triangle is used to determine the overall texture of soil based on the percentage of sand, silt, and clay

The texture of a soil can be determined from its sand, silt, and clay content using a textural triangle. The triangle above is the one created by the U.S. Department of Agriculture's Natural Resources

Conservation Service and is primarily used in the United States. Percent clay in this triangle is read on the lefthand side of the triangle, the percent silt is read on the righthand side, and the percent sand is on the bottom. For example, if a soil contains 20% clay, 40% sand, and 40% silt (total = 100%), then it is a loam.

Water

Water is the second basic component of soil. Water can make up approximately 2% to 50% of the soil volume. Water is important for transporting nutrients to growing plants and soil organisms and for facilitating both biological and chemical decomposition. Soil water availability is the capacity of a particular soil to hold water that is available for plant use.

The capacity of a soil to hold water is largely dependent on soil texture. The more small particles in soils, the more water the soil can retain. Thus, clay soils having the greatest water-holding capacity and sands the least. Additionally, organic matter also influences the water-holding capacity of soils because of organic matter's high affinity for water. The higher the percentage of organic material in soil, the higher the soil's water-holding capacity.

The point where water is held microscopically with too much energy for a plant to extract is called the "wilting coefficient" or "permanent wilting point." When water is bound so tightly to soil particles, it is not available for most plants to extract, which limits the amount of water available for plant use. Although clay can hold the most water of all soil textures, very fine micropores on clay surfaces hold water so tightly that plants have great difficulty extracting all of it. Thus, loams and silt loams are considered some of the most productive soil textures because they hold large quantities of water that is available for plants to use.

Organic Matter

Organic matter is the next basic component that is found in soils at levels of approximately 1% to 5%. Organic matter is derived from dead plants and animals and as such has a high capacity to hold onto and/or provide the essential elements and water for plant growth. Soils that are high in organic matter also have a high CEC and are, therefore, generally some of the most productive for plant growth. Organic matter also has a very high "plant available" water-holding capacity, which can enhance the growth potential of soils with poor water-holding capacity such as sand. Thus, the percent of decomposed organic matter in or on soils is often used as an indicator of a productive and fertile soil. Over time, however, prolonged decomposition of organic materials can lead it to become unavailable for plant use, creating what are known as recalcitrant carbon stores in soils.

Gases

Gases or air is the next basic component of soil. Because air can occupy the same spaces as water, it can make up approximately 2% to 50% of the soil volume. Oxygen is essential for root and microbe respiration, which helps support plant growth. Carbon dioxide and nitrogen also are important for belowground plant functions such as for nitrogen-fixing bacteria. If soils remain waterlogged (where gas is displaced by excess water), it can prevent root gas exchange leading to plant death, which is a common concern after floods.

Microorganisms

Microorganisms are the final basic element of soils, and they are found in the soil in very high numbers but make up much less than 1% of the soil volume. A common estimate is that one thimble full of topsoil may hold more than 20,000 microbial organisms. The largest of the these organisms are earthworms and nematodes and the smallest are bacteria, actinomycetes, algae, and fungi. Microorganisms are the primary decomposers of raw organic matter. Decomposers consume organic matter, water, and air to recycle raw organic matter into humus, which is rich in readily available plant nutrients.

Other specialized microorganisms such as nitrogen-fixing bacteria have symbiotic relationships with plants that allow plants to extract this essential nutrient. Such "nitrogen-fixing" plants are a major source of soil nitrogen and are essential for soil development over time. Mycorrhizae are fungal complexes that form mutalistic relationships with plant roots. The fungus grows into a plant's root, where the plant provides the fungus with sugar and, in return, the fungus provides the plant root with water and access to nutrients in the soil through its intricate web of hyphae spread throughout the soil matrix. Without microbes, a soil is essentially dead and can be limited in supporting plant growth.

Classification of Soil

Each soil possesses certain properties which develop as a consequence of the effects of a particular combination of factors affecting soil formation. The soils inherit their properties from the parent material in the initial stages of soil formation.

For example, soils developed from basic parent material are rich in basic elements and plant nutrients and are alkaline in reaction in the initials stages of development.

But as soils mature, the inherited character gradually disappears and the soils are dominated by the character acquired from the climate and vegetation. For example, as time passes, basic elements are gradually washed down and ultimately soils which had developed from basic parent materials become poor in basic elements, plant nutrients and acidic in reaction.

According to the genetic system, soils of the world have been classified into three orders: Zonal, Intra-zonal and Azonal.

Zonal Soils

The characteristics of soils belongings this order are mainly determined by the climate and vegetation. This means that soils that have developed from different parent material under the same climatic and vegetation conditions possess same general properties.

Intra-zonal Soils

General characteristics of soils belonging to this order are mainly determined by certain local conditions like poor drainage, soluble salts etc. The soil characteristics due to these conditions, dominate over the characteristics due to climate and vegetation.

Azonal Soils

Soils belongings to this order to do not possess well developed profiles because they have not been subjected to the influence of the climate and the resultant vegetation long enough for the soil profile to develop.

Zonal soils have been divided into six suborders on the basic of wider ranges of variation in climate and the resultant vegetation. Each of the sub-orders have been subdivided into great groups on the basis of narrower ranges of variation of climate and the resultant vegetation.

Intra-zonal soils have been divided into three sub-orders:

1. Hydromorphic soils that contain excessive amounts of water.

2. Holomorphic soils that contain excessive amounts of soluble salts.

3. Calciomorphic soils that contain excessive amounts of lime or gypsum.

Hydromorphic soils have been subdivided into great groups on the basis of increasing degrees of accumulation of water in soils.

Holomorphic soils have been subdivided into the two great groups, Saline and Alkali soils. Saline soils contain excessive amounts of neutral salts i.e. sulphates and chlorides of sodium, calcium and magnesium.

Alkali soils contain excessive amounts of sodium saturated clays and sodium carbonate.

Calciomorphic soils are divided into two great groups:

1. Brown forest soils that are dark brown at the surface and gradually grade into gray calcareous parent material.

2. Black earth or Grumosols that are dark brown grassland soils developed from limestone.

The order Azonal doesn't have any suborders, but the three following great groups have been recognized under it:

1. Lithosols:

   They are commonly found on steep slopes and consist of very shallow stony soils over bed rock.

2. Regosols:

   They have developed from deep soft mineral deposits. They are also very young soils like Lithosols and differ from Lithosols in that they are not stony.

3. Alluvial soils:

   The main characteristics of alluvial soils have been derived by the deposition of products of weathering of rocks by the great river systems like the Ganges, the Brahmaputra and the Indus etc. Great groups have been divided into families, each of which includes a certain number of identical soil series possessing certain common properties.

Each family has been divided into a number of series. Soils belonging to a series have originated from the same parent material, under same climate and resultant vegetation and therefore possess the same profile characteristics.

Each series has been divided into a number of soil types on the basis of the texture of the surface soil. Soil types have been divided into phases on the basis of some important deviations like slope, erosion (removal of top soil by running water or blowing wind is called soil erosion), wetness, stoniness etc.

Modern System of Soil Classification

The Genetic system of soil classification is based entirely on soil genesis. Some soils which have been cultivated for centuries have greatly changed. Yet they are still classified on the basis of the properties presumed to be possessed by them when they were not cultivated but under natural vegetation.

For these reasons, the soil survey staffs of the United States, Department of Agriculture have proposed the new system of soil classification called the seventh Approximation because it has been modified or approximated seven times.

This system of soil classification is based on the present properties of soils. However, these properties are also the consequence of soil genesis. According to the modern system of soil classification, soils of the world have been divided into ten orders based on the variation in the major soil forming processes as indicated by the presence or absence of major diagnostic Horizons.

These ten orders are as follows:

1. Entisols:

    Very young soils that do not possess a well-developed soil profile.

2. Inceptisols:

    Soil in which the horizons have just started to develop. They are relatively low in organic matter or base saturation or both and contain some weatherable minerals.

3. Aridisols:

    Soils of dry regions e.g. Deserts. They are relatively low in organic matter and possess a few pedogenic horizons.

4. Vertisols:

    Soils those are very rich in the expanding type of clay minerals and therefore have swollen and shrunk a number of times. As a consequence, they possess deep wide cracks at some stage in most of the years.

5. Mollisols:

    Grassland soils that possess a thick soft dark coloured surface horizon, relatively high in organic matter and high base saturation in all horizons.

6. Spodosols:

Soils that possess ash grey horizons near the surface due to eluviation of clay and sequi-oxide and the subsurface illuvial horizon called the spodic horizon, rich in sequioxide and humus.

7. Alfisols:

Cool humid region soils containing appreciable amounts of primary minerals from which basic elements are being slowly leached. They have gray to brown surface horizon and are medium to high in base saturation and possess an illuvial horizon of silicate clays that is more 35 per cent base saturated.

8. Ultisols:

Warm humid region, highly leached acidic soils containing low amounts of basic elements and plant nutrients. So the base saturation is low. They possess an illuvial horizon of silicate clays with base saturation lower than 35 per cent.

9. Oxisols:

Hot humid climate soils that possess an oxic horizon (this horizon contains high amounts of low cation exchange capacity clays, which are highly resistant to dispersion in water). These clays are very inactive and do not possess an illuvial horizon of silicate clays.

10. Histosols:

Organic soils that consist of plant materials. Orders (1 to 9) have been divided into sub-orders on the basis of properties associated with wetness or dryness, major parent material and vegetation, degree of decomposition of organic matter is the basis of classification of the order Histosols.

The sub orders have been divided into great groups on the basis of differentiating soil horizons and soil features. Differentiating soil horizons include those that have accumulated clay, sesquioxide and/or humus and those that have pans (hard layers). The differentiating soil features include expansion and contraction properties of clays, soil temperature and the presence of basic elements and soluble salts in the soil.

Great groups have been divided into not more than the following three subgroups:

1. The first sub-group possesses only the properties of the concerned great group.

2. The second sub-group possesses some of the properties of other subgroups, great groups, sub-orders and orders in addition to the properties of the concerned great groups.

3. The third sub-group possesses some properties that have not been found in any other great groups, sub-orders and orders in addition to the properties of the concerned great group.

Sub-groups have been divided into families on the basis of these properties important to plant growth, including soil texture, clay minerals, and soil reaction, mean annual soil temperature and mean rainfall.

Families have been divided into series on the basis of profile characteristics which means that soils belonging to a series have developed from the same parent material under the same climate and resultant vegetation and therefore possess identical profile characteristics. Soil series have been divided into phases which describe soil texture, percentage slope, stoniness, wetness, erosion, soluble salts etc.

Clay Soil

Clay soils are considered to be one of the heavier soils. The soil can hold a ton of nutrients and a lot of water due to its capillaries in between the clay particles, and it takes a little bit longer to drain. Out of all the soils, clay soil has the smallest particles.

Identifying clay soil is relatively easy to do. It is sticky when wet, it can be rolled, it can be smeared easily and it can be smoothed into a shinier finish. Clay soil has a propensity to get very hard when it dries, though, which causes the clay to crack while it is drying out.

Characteristics of Clay Soil

To the naked eye, all soil may look the same, which is why people often confuse the types of soil, but there are specific ways to differentiate them. Various soils have different colors, textures and other features that make them distinct. The reason clay soil is considered to be so difficult to work with is because it has different characteristics that set it apart from the rest.

Clay has a smooth texture because of its small particle size. Compared to other soils, a large quantity of clay can be in a small space because there are no gaps, whereas soil with large particles has way more gaps. If the soil has large particles, this will give it a rougher texture, while the small particles in clay give it a smoother texture.

Due to clay's small particle size, the soil is very dense, and it bonds together, which is why some plant roots can't penetrate clay soil. A positive to this, though, is that clay soils are resistant to erosion. A negative is that it takes longer for clay-heavy soil to warm up after a season of cold weather.

Different Percentages of Clay Soil

There are four types of clay soil that differ in characteristics depending on the amount of clay in the soil. The different percentages of clay soil include silt soils which have 0 to 10 percent clay, clay soils with 10 to 25 percent clay, clay soils with 25 to 40 percent clay and clay soils with 40 percent clay.

1. Silt Soils With 0 to 10 Percent Clay

This type of soil has a propensity to form a crust, which makes the soil hard. You cannot overtill this soil because it will become very compact, and water will not be able to permeate the soil. This type of soil is generally easy to till, but you should not till this type of soil in wet conditions.

2. Clay Soils With 10 to 25 Percent Clay

Crusting with this type of soil can be really serious. It is difficult to till this type of clay soil because it has a low amount of clay and organic material.

3. Clay Soils With 25 to 40 Percent Clay

This type of soil is dark in color and must be tilled with the correct water content to be cultivated correctly. If the environment runs very dry, the clay can clod.

4. Clay Soils With 40 Percent Clay

Heavy clay soils need a lot of recompacting in order to be considered an optimal soil. It is very nutrient rich, but it cannot be tilled in wet conditions.

Properties of Clay Soil

Though different soils have a wide range of colors, textures and other distinguishing features, there are only three types of soil particles that geologists consider distinct. The quality of soil depends on the amount of sand, loam and clay that it contains, because soils with differing amounts of these particles often have very different characteristics. Soil with a large amount of clay is sometimes hard to work with, due to some of clay's characteristics.

Particle Size

Clay has the smallest particle size of any soil type, with individual particles being so small that they can only be viewed by an electron microscope. This allows a large quantity of clay particles to exist in a relatively small space, without the gaps that would normally be present between larger soil particles. This feature plays a large part in clay's smooth texture, because the individual particles are too small to create a rough surface in the clay.

Structure

Because of the small particle size of clay soils, the structure of clay-heavy soil tends to be very dense. The particles typically bond together, creating a mass of clay that can be hard for plant roots to penetrate. This density is responsible for clay-heavy soil being thicker and heavier than other soil types, and clay soil takes longer to warm up after periods of cold weather. This density also makes clay soils more resistant to erosion than sand or loam-based soils.

Organic Content

Clay contains very little organic material; you often need to add amendments if you wish to grow plants in clay-heavy soil. Without added organic material, clay-heavy soil typically lacks the nutrients and micronutrients essential for plant growth and photosynthesis. Mineral-heavy clay soils may be alkaline in nature, resulting in the need for additional amendments to balance the soil's pH before planting anything that prefers a neutral pH. It's important to test clay-heavy soil before planting to determine both the soil's pH and whether it lacks important nutrients such as nitrogen, phosphorus and potassium.

Permeablity and Water-Holding Capacity

One of the problems with clay soil is its slow permeability resulting in a very large water-holding capacity. Because the soil particles are small and close together, it takes water much longer to move through clay soil than it does with other soil types. Clay particles then absorb this water, expanding as they do so and further slowing the flow of water through the soil. This not only prevents water from penetrating deep into the soil but can also damage plant roots as the soil particles expand.

Identifying Clay

There are several tests you can use to identify clay soils. If rubbed between your fingers, a sample of clay soil often feels slick and may stick to your fingers or leave streaks on your skin. Rubbed clay soil often takes on a shiny appearance as well, as opposed to the rough texture you would see with other soils. Clay soils do not crumble well, and a sample of clay can typically be stretched slightly without breaking. When wet, clay soils become slick and sticky; the soil may also allow water to pool briefly before absorption due to the slow permeation. Visually, clay soils seem solid with no clear particles, and may have a distinct red or brown color when compared to the surrounding soil.

Sandy Soil

Sandy soil is often called 'light soils' because they are relatively 'light' or easy to work with when it comes to ploughing, planting and cultivating. But they're certainly not 'light' or easy to manage! These soils tend to dry out quickly. However, some sandy soils lie on a rocky layer and so can become waterlogged after a lot of rain. In this case, you'll need drainage trenches. Other sandy soils have a clay or loam base, which is only discovered when digging holes to determine the soil profile.

Such soils can be highly productive as the heavier soil below can trap leached nutrients and hold moisture. Sandy soils have very little clay to retain nutrients and so are not fertile. Growing vegetables in sand is similar to farming with hydroponics, where the crops are planted in gravel or some other medium and all the nutrients are supplied via the irrigation water. With hydroponic farming, however, the water containing the nutrients is circulated. In sandy soil, the nutrients are washed through the soil and mostly lost. To manage this, the golden rule is 'less, more often'. Because sand dries out quickly, apply less water more often. With fertiliser, too, give lighter, more frequent applications to allow the roots to take up the nutrients before these are lost.

Organic fertiliser, such as compost, helps with this process. This 'holds' any other fertilisers that you apply. Then, as the compost decomposes, it gradually releases nutrients through the activity of soil organisms. This is called 'mineralisation', and it's usually the most practical way to fertilise sandy soils.

In general, it's difficult to build up the organic content of sandy soils, especially in warm areas. To achieve the best results, apply the golden rule: smaller quantities of organic matter more often. Grass mulch also helps to fertilise the soil, as well as keeping it cool in hot weather and reducing weeds.

Loss of Nutrients

Leaching of nitrogen can cause serious health problems for your plants, and should be addressed with organic fertiliser. Loss of calcium can also produce poor results as it causes acid soils, especially in high-rainfall areas; apply agricultural lime to these soils from time to time. Your fertiliser agent or extension officer can advise you on the quantity needed. Organic fertilisation also helps to produce better crops in acid soils.

Eelworm can be a problem in sandy soils, so it's a good idea to rotate crops with crops that are not susceptible to this serious pest. Despite these difficulties, all vegetables can be grown in sandy soil. In fact, this type of soil can be extremely productive if you use the right management methods.

Coming to Terms with Soil

Simply put, soil, the medium in which rooted plants grow, is a mixture of solids, water and air. The solids include minerals, tiny rock particles, organic matter from animals and plants, and minute living organisms.

There are three basic soil types – sandy, loam and clay. However, most lands have a variable mixture of these, such as 'sandy loam' – soil that's part sand, part loam. The ideal way to grow vegetables is to conduct market research to decide on what crops are in demand, then find a piece of land with the best possible soil for the crop. In the real world, however, most farmers and gardeners have to learn to work with the soil they have and make the most of it.

Drainage Ditches

The most efficient drainage system – one that allows you to drain the biggest area – is the fishbone pattern. As the name suggests, this consists of channels joining a central ditch at about a 45° angle.

The number of channels and the length of the ditch will be determined by the size of the area under cultivation. Make sure the channels and ditch are angled correctly to carry the water in the right direction.

The channels which guide the water into the central ditch and the ditch itself are filled with rocks, covered with plastic (old fertiliser bags can be used), and filled with soil. This makes the system almost invisible and you can cultivate the land as normal, even over the channels. You can even build a dam to collect the water that flows out of the central drainage ditch.

Soil Profile

If you dig a deep hole in your soil, you'll notice various layers. These so-called 'horizons' differ in texture, structure, colour, ability to hold water and so on, and make up the soil profile.

The surface layer, which contains grass or plant life, is the 'O' horizon. The two layers immediately below this – the 'A' and 'B' horizons – are regarded as the 'true soil', as most of the chemical and biological activity that helps plants grow takes place there.

Sandy Soil Improvement

The best sandy soil amendments are ones that increase the ability of the sandy soil to retain water and increase the nutrients in the soil as well. Amending sandy soil with well rotted manure or compost (including grass clippings, humus and leaf mold) will help to improve the soil the fastest. You can also add vermiculite or peat as sandy soil amendments, but these amendments will only add to the soil's ability to hold onto water and will not add much nutrient value to the sandy soil.

When amending sandy soil, you need to watch the salt levels of the soil. While compost and manure are the best way to amend sandy soil, they contain high levels of salt that can stay in the soil and damage growing plants if the salt level builds up too high. If your sandy soil is already high in salt, such as in a seaside garden, be sure to use plant only based compost or sphagnum peat, as these amendments have the lowest salt levels.

Uses of Sandy Soil

The uses of sandy soil in different sectors are numerous. Followings are the uses of sandy soil:

- Agricultural need
- Easy drainage
- Construction
- Foundation
- Beauty
- Frictional properties
- Low settlements
- Changing pH
- Others

A brief description of these uses of sandy soil is given below:

Agricultural Need

Sandy soil is usually dry, nutrient and fast draining. It is used for plowing, planting and cultivating. The useful vegetables like potatoes, grams, tomatoes etc require a minimal percentage of soil for a specific period. The percentage varies from vegetable to vegetable. Sandy soil also provides a good ground for farmers to collect falling nuts.

Easy Drainage

Sandy soil has great drainage properties. It drains easily and quickly. It is used to improve soil drainage. The interesting feature is in a land full of sandy soil, work can be done right after a rain even if it's heavy without any difficulty. It filters water in big deposits through which water is shred and recollected through the channels at the bottom. Sandy soil also retains flower water like soaking the water.

Construction

Sandy soil doesn't get sticky. It is cohesionless. It has light and loose structure. That's why it can be easily used for construction purpose. Sandy soil can be a great aggregate for concrete. Also, it can be used as a construction material of mortar aside cement. Sandy soil is also used for the erection of exterior rendering materials. It is used because of its chemical resistance. Sandy soil also can be used the best as filling sand.

Foundation

Sandy soil provides a base to foundations on swampy ground. Densification below foundation is not required as the soil is naturally in a dense state. Sandy soil provides a reliable foundation for which the structure that relies on this soil has a greater degree of reliance than these built-in other soils. Sandy soil is maneuvered for the improvement of ground with the use of soil replacement method. It is used to replace soft clays of foundation to improve bearing capacity of the soil.

Beauty

Sandy soil starts from the almost sandy surface like beaches. It increases the beauty of the beach. Also, it's used in gardening and kids playgrounds for safety by providing a soothing context.

Frictional Properties

Sandy soil has very good frictional properties. The frictional properties are used in the construction of reinforced soil structure with geosynthetics reinforcement.

Low Settlements

Sandy soils have low settlements as it does not undergo consolidation with time. Moreover, it has immediate settlements.

Changing pH

The pH level of sandy soil can easily change the pH level of soil like clay. The pH level of sandy soil is between 7.00 and 8.00.

Others

Sandy oil is used to reduce the velocity of water. It also helps to percolate the soil and raise the water table.

Management of Sandy Soils

Farmers preparing sandy soil for planting acacia tree seedlings, Senegal

Sandy soils are those that are generally coarse textured until 50 cm depth and consequently retain few nutrients and have a low water holding capacity.

Soil management practices which lead to an increase in the fine fraction are helpful in improving soil properties and crop productivity.

- Fertilization of these soils is considered essential. Inorganic fertilization is the main practice.

- Application of organic manures can supply nutrients in slowly available forms and improve soil physico–chemical properties.

- Surface application of organic manures to sandy soils does not last so the manure should be dug deeper into the soil or a carpet-like layer spread of not less than one centimeter thick, which will improve water storage, biological activity, nutrient status and increase yields.

- Mulch can be added to improved water storage by reducing evaporation. Crop residues, on the surface of the soil reduce evaporation losses, decrease the range between maximum and minimum soil temperature, and reduce wind erosion.

- For tillage to be really effective, it has to be done at the earliest possible time after irrigation or rainfall when the evaporation rate is still high.

- Minimum tillage, maintenance of a cover crop, strip cropping, crop rotations, control of grazing and establishment of shelter belts and windbreaks are some of the protective measures to counter the high susceptibility of sandy soils to erosion.

- Besides the conventional dry vegetation method, use of artificial surface sealants such as petroleum, synthetic rubber, chemicals and water soluble plastics have also been adopted for dune and drift sand stabilization.

- Afforestation with selected trees and shrubs is a complementary measure that should follow stabilization of dunes.

- Overgrazing on coarse textured soils must be avoided. The introduction of rotational grazing helps to combat this hazard. It might be better not to permit grazing but to use fodder cut on feeding lots.

Silt

Silt is granular material of a size between sand and clay, whose mineral origin is quartz and feldspar. Silt may occur as a soil (often mixed with sand or clay) or as sediment mixed in suspension with water (also known as a suspended load) and soil in a body of water such as a river. It may also exist as soil deposited at the bottom of a water body, like mudflows from landslides. Silt has a moderate specific area with a typically non-sticky, plastic feel. Silt usually has a floury feel when dry, and a slippery feel when wet. Silt can be visually observed with a hand lens, exhibiting a sparkly appearance. It also can be felt by the tongue as granular when placed on the front teeth (even when mixed with clay particles).

Sources

Silt is created by a variety of physical processes capable of splitting the generally sand-sized quartz crystals of primary rocks by exploiting deficiencies in their lattice. These involve chemical weathering of rock and regolith, and a number of physical weathering processes such as frost shattering and haloclasty. The main process is abrasion through transport, including fluvial comminution, aeolian attrition and glacial grinding. It is in semi-arid environments that substantial quantities of silt are produced. Silt is sometimes known as "rock flour" or "stone dust", especially when produced by glacial action. Mineralogically, silt is composed mainly of quartz and feldspar. Sedimentary rock composed mainly of silt is known as siltstone. Liquefaction created by a strong earthquake is silt suspended in water that is hydrodynamically forced up from below ground level.

Grain Size Criteria

In the Udden–Wentworth scale (due to Krumbein), silt particles range between 0.0039 and 0.0625 mm, larger than clay but smaller than sand particles. ISO 14688 grades silts between 0.002 mm and 0.063 mm (sub-divided up into three grades fine, medium and coarse 0.002 mm to 0.006 mm to 0.020 mm to 0.063 mm). In actuality, silt is chemically distinct from clay, and unlike clay, grains of silt are approximately the same size in all dimensions; furthermore, their size ranges overlap. Clays are formed from thin plate-shaped particles held together by electrostatic forces, so present a cohesion. Pure silts are not cohesive. According to the U.S. Department of Agriculture Soil Texture Classification system, the sand–silt distinction is made at the 0.05 mm particle size.

The USDA system has been adopted by the Food and Agriculture Organization (FAO). In the Unified Soil Classification System (USCS) and the AASHTO Soil Classification system, the sand–silt distinction is made at the 0.075 mm particle size (i.e., material passing the #200 sieve). Silts and clays are distinguished mechanically by their plasticity.

Environmental Impacts

A silted lake located in Eichhorst, Germany

Silt is easily transported in water or other liquid and is fine enough to be carried long distances by air in the form of dust. Thick deposits of silty material resulting from deposition by aeolian processes are often called loess. Silt and clay contribute to turbidity in water. Silt is transported by streams or by water currents in the ocean. When silt appears as a pollutant in water the phenomenon is known as siltation.

Silt, deposited by annual floods along the Nile River, created the rich, fertile soil that sustained the Ancient Egyptian civilization. Silt deposited by the Mississippi River throughout the 20th century has decreased due to a system of levees, contributing to the disappearance of protective wetlands and barrier islands in the delta region surrounding New Orleans.

In southeast Bangladesh, in the Noakhali district, cross dams were built in the 1960s whereby silt gradually started forming new land called "chars". The district of Noakhali has gained more than 73 square kilometres (28 sq mi) of land in the past 50 years.

With Dutch funding, the Bangladeshi government began to help develop older chars in the late 1970s, and the effort has since become a multi-agency operation building roads, culverts, embankments, cyclone shelters, toilets and ponds, as well as distributing land to settlers. By fall 2010, the program will have allotted some 100 square kilometres (20,000 acres) to 21,000 families.

A main source of silt in urban rivers is disturbance of soil by construction activity. A main source in rural rivers is erosion from plowing of farm fields, clearcutting or slash and burn treatment of forests.

Types of Plants Grown in Silty Soil

Silty soil has characteristics of fine particle size, prone to compaction and moisture retention but without the drainage problems typical of clay soil. It is usually found in areas once covered by water or areas near water such as riverbeds, deltas and lakes. Plants that grow well in clay soil will thrive in silty soil. The added drainage, high nutrient content and stable base of silt makes it suitable for growing a variety of plants, including herbaceous perennials, roses and other shrubs, bulb plants and ferns.

Perennials

- Hostas (*Hosta* spp.) typically require shade and thrive in damp soil, which makes them options for silty soil. Commonly grown for their foliage, over 40 varieties are available. Hostas are perennial, or hardy, in U.S. Department of Agriculture plant hardiness zones 3 through 9, depending on their variety.

- Hellebore *(Helleborus x hybridus, Ballard's Group)*, hardy in USDA zones 4 through 9, is a group of flowering perennials well-suited for the moist, well-draining conditions presented by silty soil. Their showy flowers appear in early spring.

- Cranesbill (*Geranium* spp.), flowering perennials also called hardy geraniums, grow in moist soil that drains well, making them well-suited for silty gardens. They are hardy in USDA zones 5 through 8, depending on the type.

Roses

(_Rosa _spp.) grow well in silt because they prefer soil on the heavy side. Hundreds of rose varieties are available. They are hardy in USDA zones 2 through 11, depending on their kind.

- 'Polar Ice' rose (Rosa rugosa 'Polar Ice') is hardy in USDA zones 2 through 9.

- Hedgehog rose (*Rosa rugosa* var. *alba*), also hardy in USDA zones 2 through 9, works well in cottage gardens. It features simple open flowers.

- Lady Banks rose (*Rosa banksia 'Lutea'*), with pale-yellow flowers, is hardy in USDA zones 7 through 11.

Other Shrubs

- Butterfly bush (Buddleja davidii), hardy in USDA zones 5 through 9, grows well in silty soil because it adapts to wet and dry conditions. This shrub is so successful that it is invasive in some areas; keep it pruned and remove unwanted seedlings to prevent that problem.

- Japanese barberry (Berberis thunbergii), hardy in USDA zones 4 through 10, grows well in silt soil. With many cultivars to choose from, you should be able to find one that is just right for your landscape.

- Smoke tree (Cotinus coggygria), hardy in USDA zones 5 through 8, makes a statement in the landscape with its smokelike foliage. It grows well in silty soil that has good drainage.

Ferns

A variety of ferns grow well in the moist, often wet conditions of silty soil.

- Male fern (*Dryopteris filix-mas*) is a 2- to 3-foot-tall plant for shady areas. It is hardy in USDA zones 4 through 8.

- Ostrich fern (*Matteuccia struthiopteris*) grows 3 to 6 feet tall and prefers moist, shady areas. It is hardy in USDA zones 3 through 7.

Bulbs

Some flowering bulbs are well-suited for silty soil. Their bloom times vary.

- Snowdrop (*Galanthus nivalis*), hardy in USDA zones 3 through 7, blooms in late winter or spring.

- Daffodils (*Narcissus* spp_._), hardy in USDA zones 4 through 8, flower in early spring or in the middle of spring.

- Crocus (*Crocus vernus*), hardy in USDA zones 3 through 8, flower in early spring.

- Snowflake (*Leucojum aestivum*), hardy in USDA zones 4 through 8, blooms in early to mid-spring.

Peat Soil

Peat soils are formed from partially decomposed plant material under anaerobic water saturated conditions. They are found in peatlands (also called bogs or mires). Peatlands cover about 3% of the earth's land mass; they are found in the temperate (northern europe and america) and tropical regions (south east asia, south america, south africa and the caribbean).

Peat soils are classified as histosols. These are soils high in organic matter content. Peat formation is influenced by moisture and temperature. In highly saturated anaerobic soils, decomposition of plant material by micro organisms is slowed down, resulting in high carbon accumulation. In colder climates decomposition of plant material by micro organisms is slowed down leading to quicker peat formation. The carbon content of peat soils makes peatland a major storage of carbon on the earth surface. This is why its importance in fighting climate change can never be overemphasized.

Some Economic Benefits of Peatlands

Peatlands bring enormous economic benefits to regions where they are found.

1. Peat is extracted for use as horticultural compost. It is highly sought after in commercial horticulture because of its high water retaining ability and flow of air.

2. Peat is used for fuel to generate electricity. It is also sold as briquettes for heating homes in cold climatic regions.

3. Peatlands are drained and used for agricultural purposes (pasture and crop production) and forestry.

Peat use for forestry and agriculture are beneficial but it alters the natural peatland hydrology. This causes oxidation of stored carbon therefore declining its organic matter content. During peat extraction, peat is drained and dried before storage or transportation for sale. These processes reduce the water content and encourage microbial decomposition of organic matter. The result of this is the release of greenhouse gasses such as CO_2 and N_2O.

Consequences of Peat Disturbance

Apart from greenhouse gas emission, peatland disturbance brings a number of other changes:

1. Drainage of peatland causes decline in biodiversity because its natural hydrological habitat is disturbed. Peatlands provide habitation for diverse species of meadow birds, animals, vegetation and insects.

2. Peat oxidation can lead to release of dissolved organic matter and peat particles into surface waters.

3. Peat oxidation can lead to the loss of a historical heritage. Peat soils have the ability to store human remains or ancient artefacts for thousands of years; since they have very minimal microbial decomposition. A good example of this is the 4000 year old body of a man found in peat from Cashel-Central Ireland.

4. Peat soils drained for agricultural purposes are more vulnerable to wind and water erosion when the topsoil is severely dry.

5. Drainage of peatland can lead to peat fires which destroy forestland and habitation and further increase the emission of CO_2 to the atmosphere.

Although the above listed are negative consequences of peat disturbance, it is good to acknowledge that when peat soils are drained for use in agriculture, decomposition of organic matter is accelerated leading to the mineralization of nitrogen (a vital nutrient for plant growth). This in fact is a good thing. You can learn more about this from the nitrogen cycle.

Peatland Conservation and Restoration

1. Conserve wet peatlands: This approach is preventive and avoids the expensive cost of restoring peatlands to their natural hydrological state. This is simply putting a stop to the drainage of peatlands. There is no need for soil restoration projects if efforts are made to keep the soil in its natural state. People in surrounding communities must be educated on the benefits of conserving the peatland natural ecosystem.

2. Use of paludiculture: This method also maintains the wetness of peatlands. Paludiculture involves the cultivation of biomass crops on peatlands without disturbing the peat natural hydrology or ecosystem. There is no drainage of the peatland involved. It has multiple benefits; it reduces peat oxidation, greenhouse gas emission and at the same time supplies biomass used for combustion. It also serves as a source of food for neighbouring communities as some edible crops grow on the wetland. Examples of edible crops in paludiculture are wild rice -Zizania aquatica (also called floating rice), wild edible berries (blue berries, black currant, black raspberries) and sweet grass Hierochloe odorata. Paludiculture is practiced in Europe (Russia and Belarus) North America and Asia (Indonesia and Malaysia).

3. Re-wetting and restoration: In re-wetting effort is made to restore the soil back to its natural hydrological, anaerobic state by raising the water table level to the land surface. Re-wetting reduces CO_2 and N_2O emissions but increases the emission of CH_4 (which is released naturally in undisturbed peatlands). Under anaerobic conditions decomposition of plant material by micro organism is slow but still in action. Decomposition of organic material under this condition is carried out by methanogenic Archaea (a methane producing micro organism).

Loam Soil

Three layers of subsurface loam; surface layer is dark brown fine sandy loam, subsurface layer is pale brown fine sandy loam, subsoil is red clay loam and sandy clay loam.

Loam is soil composed mostly of sand (particle size > 63 micrometres (0.0025 in)), silt (particle size > 2 micrometres (7.9×10^{-5} in)), and a smaller amount of clay (particle size < 2 micrometres (7.9×10^{-5} in)). By weight, its mineral composition is about 40–40–20% concentration of sand–silt–clay, respectively. These proportions can vary to a degree, however, and result in different types of loam soils: sandy loam, silty loam, clay loam, sandy clay loam, silty clay loam, and loam.

In the USDA textural classification triangle, the only soil that is not predominantly sand, silt, or clay is called "loam". Loam soils generally contain more nutrients, moisture, and humus than sandy soils, have better drainage and infiltration of water and air than silt and clay-rich soils, and are easier to till than clay soils. The different types of loam soils each have slightly different characteristics, with some draining liquids more efficiently than others. The soil's texture, especially its ability to retain nutrients and water are crucial. Loam soil is suitable for growing most plant varieties.

Bricks made of loam, mud, sand, and water, with an added binding material such as rice husks or straw, have been used in construction since ancient times.

Use in Farming

Fine, loam-rich field ideal for farming vegetables in the UK

Loam is considered ideal for gardening and agricultural uses because it retains nutrients well and retains water while still allowing excess water to drain away. A soil dominated by one or two of the three particle size groups can behave like loam if it has a strong granular structure, promoted by a high content of organic matter. However, a soil that meets the textural definition of loam can lose its characteristic desirable qualities when it is compacted, depleted of organic matter, or has clay dispersed throughout its fine-earth fraction.

Loam is found in a majority of successful farms in regions around the world known for their fertile land. Loam soil feels soft and crumbly and is easy to work over a wide range of moisture conditions.

Use in House Construction

Loam may be used for the construction of houses, for example in loam post and beam construction. Building crews can build a layer of loam on the inside of walls, which can help to control air humidity. Loam, combined with straw, can be used as a rough construction material to build walls. This is one of the oldest technologies for house construction in the world. Within this there are two broad methods: the use of rammed earth, or unfired bricks (adobe).

Loam with timber framing

House with loam ground floor

Loam-timber-framed 1707 house, under restoration in Brandenburg

Great Mosque of Djenné

Saline and Alkaline Soil

The presence of an excess of sodium salts and the predominance of sodium in the exchangeable complex are divided into the two main groups:

1. Saline soils and

2. Alkaline soils.

- Saline Soils

  Saline soils contain an excess of sodium salts, but its colloidal material is not yet sodium-ised.

- Alkali Soils

  In the case of alkali soils, the exchange complex contains appreciable quantities of exchangeable sodium. Such soils may or may not contain excess salts.

Alkali soils may be divided into following groups:

- Saline-alkali soils

  When they contain soluble salts in excess they are known as saline-alkali soils.

- Non-saline-alkali soils (Alkali soil)

  When they do not contain soluble salts, they are called non-saline-alkali soils.

- Degraded alkali soils

  Under certain circumstances the clay complex of some alkali soils is broken down to give rise to degraded alkali soils.

The various types of alkaline soils are shown diagrammatically as under:

Table: Characteristics of salt affected soils (Saline and Alkaline soils).

Nature of Soil	Soil characteristics		ESP
	Ph	EC	
Saline	< 8.5	> 4 mmhos/cm	< 15
Alkali	> 8.5	< 4 mmhos/cm	< 15
Saline- Alkali	8.5	> 4 mmhos/cm	< 15

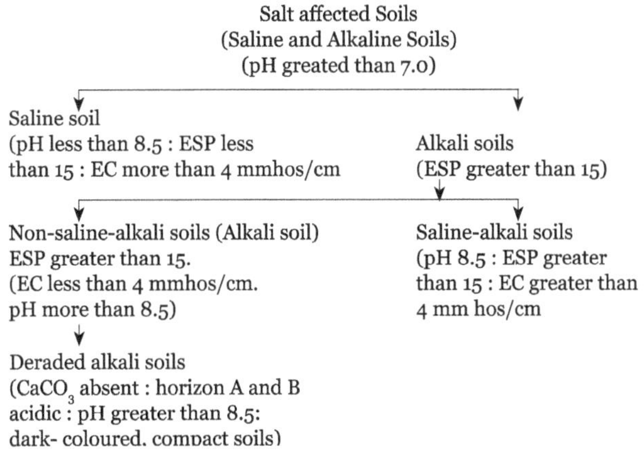

Salt affected Soils
(Saline and Alkaline Soils)
(pH greated than 7.0)

Saline soil
(pH less than 8.5 : ESP less
than 15 : EC more than 4 mmhos/cm

Alkali soils
(ESP greater than 15)

Non-saline-alkali soils (Alkali soil)
ESP greater than 15.
(EC less than 4 mmhos/cm.
pH more than 8.5)

Saline-alkali soils
(pH 8.5 : ESP greater
than 15 : EC greater than
4 mm hos/cm

Deraded alkali soils
($CaCO_3$ absent : horizon A and B
acidic : pH greater than 8.5:
dark- coloured. compact soils)

Table: Area under saline and alkaline soils (lakh hectares).

State	Area	State	Area
1. U.P.	12.95	9. Karnataka	4.04
2. Gujarat	12.14	10. M.P.	2.24
3. W.Bengal	8.50	11. A.P.	0.42
4. Rajasthan	7.28	12. Delhi	0.16
5. Punjab	6.89	13. Kerala	0.04
6. Maharashtra	5.34	14. Bihar	0.04
7. Haryana	5.26	15. Tamil Nadu	0.04
8. Orissa	4.04	Total	69.49
			or
			7.00 million hectates

Characteristics of Saline and Alkaline Soils

Saline Soil

When the soil contains excess of sodium salts and clay complex still contains exchangeable calcium, the soil is known as saline soil or white alkali or brown alkali soil. The process of accumulation of salts leading to the formation of soils is known as salinization.

1. Saline soils contain usually chloride, sulphate, bicarbonates and sometime nitrates of sodium. The presence of chloride and sulphate of sodium gives a white colour on the soil surface. When nitrates are in excess they give a brown colour to the soil.

2. Exchangeable sodium percentage (ESP) is very low, being less than 15% of the total cation exchange capacity (C.E.C.).

3. As a consequence of low ESP, generally pH varies between 7.5 and 8.5.

4. Total soluble salt content is more than 0.1%. it is high enough to interfere with normal growth of most plant species.

5. Electrical conductivity (E.C.) of solution extract (saturated soil) is 4 or more m mhos/cm.

6. Saline soils remain in a flocculated condition (granulated). It is permeable to water and air.

7. Saline soils usually have a surface crust of white salts, especially in the season when the net movement of soil moisture is upward. Salts dissolved in the soil water move up to the surface, where they are left as a crust when the water evaporates.

Alkaline Soil (Sodic Soil)

(a) Non-saline-alkali Soils

The characteristic features are the presence of collodial complex that is saturated with exchangeable sodium, and the absence of appreciable quantities of soluble salts. These soils are often called 'black alkali' soils, because they are black, owing to the effect of the high sodium content which causes the dispersion of the organic matter. These soils are also called typical usarsoils. These soils contain sodium carbonates ($Na_2 CO_3$) in abundance.

- Exchangeable sodium percentage is greater than 15%.

- Consequently pH ranges from 8.5 to 10 (strongly alkaline).

- Total soluble salt (sodium) content is less than 0.15.

- Electrical conductivity (EC) is usually less than 4 mmhos/cm.

- Colloidal complex is deflocculated and dispersed. The clay swells and chokes the soil pores. Hence, permeability to water and air is poor (or infiltration and aeration is slow).

- The presence of free sodium carbonate has a toxic effect on plant roots. Also, the high pH and poor physical condition of soil adversely affect plant growth.

- Sodium carbonate absorbs organic matter, so there is great depletion of organic matter. Therefore, these soils are almost barren (Usar).

(b) Saline-alkali Soils

These soils are both saline and alkali. There can be all stages in transition with varying degree of dominance of salt content and pH. According to movement of soluble salts, formation of saline-alkali and non-saline alkali soils depends. Soil contains Na-clay as well as excess soluble salts.

If the soluble sodium salts are not leached out due to the insufficiency of rain water, they remain in the soil. The soil thus contains Na-clay and excess soluble, salts in solution. Such soils are known as saline-alkali soils. They are thus, developed as a result of the combined process of salinization and alkalization. In spite of the presence of sodium clay (Na-clay) the soil remains friable and possesses aggregate (flocculated). This is because the presence of sodium salts does not allow the sodium clay to get dispersed and keeps it flocculated.

Thus, this soil behaves more or less like saline soils. If due to much water soluble salts are leached down, and soil contains Na-clay only. Thus, this soil behaves more or less as non-saline-alkali soil. Therefore, the soil structure becomes un-favourable for the entry and movement of air and water.

Usually these soils have the following characteristics:

1. Exchangeable sodium is more than 15%.

2. A variable pH, usually above 8.5, depending upon the relative amounts of exchangeable sodium and soluble salts. When soluble salts are leached downward, the pH will rise above 8.5, but when the soluble salts again accumulated, the pH again falls to 8.5.

3. Generally soluble salts content is more than 0.1%.

4. Electrical conductivity is greater than 4 mmhos/cm.

Degraded Alkali Soils

The soil does not contain free calcium carbonate ($Ca\ CO_3$). As a result of prolong leaching under this condition, Na-clay hydrolyses NaOH which combines with CO_2 or soil air and forms sodium carbonate (Alkaline condition).

$$Na\text{-}clay + H_2O \rightleftharpoons \underset{(acid\ soil)}{H\text{-}clay} + NaOH$$

$$2\,NaOH + CO_2 \rightarrow \underset{(alkali\ soil)}{Na_2\,CO_3} + H_2O$$

Sodium carbonate ($Na_2\ CO_3$) dissolves humus. Humus (organic matter) is deposited in the lower layer. The lower layer thus, acquires a black colour. At the same time, a part of exchangeable sodium of the surface layer is replaced by hydrogen. H-clay (acid soil) formed in this way does not remain stable. The process of break-down of H-clay under alkaline condition is known as solodization and the soil as formed is called Solod, Soloth or degraded alkali soil.

1. The soil reaction of the surface layer is acidic (pH 6.0). This layer is usually very thin, hardly a few inches in depth.

2. The lower layer which constitutes the main soil body has a high pH (more than 8.5).

3. ESP is greater than 15%.

4. EC less than 4 mmhos/cm.

5. The lower layer has black colour.

6. It develops columnar (prism-like) structure.

7. Soils become compact and has low infiltration, and permeability.

Formation of Saline and Alkaline Soil

Origin or development of saline and alkaline soil depends upon following factors:

(i) Arid and Semi-Arid Climate

Alkaline soils are those that have an alkaline reaction or whose pH is greater than 7.0. Alkalinity is due to sodium salts in soil solution or the presence of sodium clay or both. They are formed in arid and semi-arid regions which have very low rainfall and high evaporation.

The low rainfall in these regions is not sufficient to leach out the soluble products of weathering and hence, the salts accumulate in the soil. During rain, the salts dissolve in rain water and move down in the lower layers. However, due to the limited rainfall, the downward movement is restricted to a short distance only. In dry weather, the salts move up with the water and are brought up to the surface where they are deposited as the water evaporates.

(ii) Poor Drainage of Soil

During the periods of high rainfall, the salts are leached from the upper layer and, if the drainage is impeded, they accumulate in the lower layer. When water evaporates, the salt is left in the soil. Such soils are generally developed in low-lying areas or in basin shaped areas.

(iii) High Water Table

The ground waters of arid regions usually contain considerable quantities of soluble salts. If the water table is high, large amounts of water move to the surface by capillary action and the evaporated, leaving soluble salts on the surface.

(iv) Overflow of Sea Water over Lands

Low lying areas near the sea which get sea water during tides. Salt water accumulates and enrich the soils with salts.

(v) Introduction of Irrigation Water

The ground water of arid regions are generally saline in nature. With injudicious irrigation the percolating water may get linked with the saline ground water. During dry weather the soluble salts of the ground water may, thus, get carried to the surface and increase the salinity of the land. The irrigation water may be itself rich in soluble salts and add to the salinity of the soils.

(vi) Salts Blown by Wind

In arid regions near the sea, lot of salt is blown by wind year after year and get deposited on the lands. Due to low rainfall they are not washed back to sea and thus, add salinity to the land. The salinity of Rajasthan has developed to a great extent, due to this reason.

(vii) Saline Nature of Parent Rock Materials

If soils develop from saline nature of parent rock materials, soil would be saline.

(viii) Excessive Use of Basic Fertilizers

Use of alkaline fertilizers like sodium nitrate, basic slag etc., may develop alkalinity in soil.

(ix) Humid and Semi-Humid Regions

Alkaline soils develop in other areas also, e.g., in semi-humid and temperate regions, especially in depressions where drainage is defective and where the underground water table is high or close to the surface. There are three distinct stage in the evolution of saline and alkali soils.

They are as follows:

1. Saline soils (Salinization):

Soil contains excess of sodium salts while the clay- complex (soil-colloid) still contains exchangeable calcium and magnesium. In these soils the colloids are not damaged by sodium.

2. Saline-alkali soils:

When soluble sodium salts accumulate in a soil over a prolong period, form sodium clay (sodium becomes the predominant cation in soil solution). If the soluble salts (sodium) are not leached out due to the insufficiency of rain, they remain in the soil. They are thus, developed as a result of the combined process of salinization and alkalization. Sodium salts keep soils in flocculated conditions.

$$Ca-clay + Na-salts \rightleftharpoons Na-clay+ Ca-salts$$

3. Alkalinization (non-saline-alkali soils):

When soluble salts (from saline-alkali soils) are removed by leaching as a result of the increase in rainfall, it gives rise to non- saline-alkali soil (only Na-clay in the soil colloids). Calcium carbonate (Ca CO_3) reacts with Na-clay and give rise to Ca-clay and sodium carbonate (Na$_2$ CO_3). Duc to low $CaCO_3$, Na$_2$ CO_3 converts Ca-clay into Na-clay. The clay is thus sodium saturated.

$$Na-clay + CaCO_3 \rightleftharpoons Ca-clay+ Na_2\ CO_3$$

If $CaCO_3$ is absent, it forms degraded alkali soils. Na-clay hydrolyses (during leaching) and liberates NaOH which combines with the CO_2 and forms sodium carbonate.

$$Na-clay + H_2O \rightleftharpoons H\text{-} clay(acid) + NaOH$$
$$2\,NaOH+ CO_2 \rightleftharpoons Na_2\ CO_3+ H_2O$$

Detrimental Effects of Soil Salinity and Alkalinity

Saline Soils

(i) Absorption of Water and Nutrients

Excessive salts in the soil solution increase the osmotic pressure of soil solution in comparison to cell sap. This prevents absorption of moisture and nutrients in adequate amounts by the roots.

(ii) Salt toxicity:

When the concentration of soluble salts increase to high level then it produces toxic effect directly to plants. Saline soils are usually barren but potentially productive soils.

Alkaline Soil

(i) Dispersion of soil particles

Under alkali soil conditions, the damage is not due to salt concentration. The sodium adsorbed by clay and colloids causes dispersion of clay which results in a loss of desirable structure and development of compact soil.

(ii) Physical properties affected

Due to compactness of soil, aeration, permeability, drainage and microbiological activity are reduced.

(iii) Availability of plant nutrients reduced

The high pH in alkali soil causes a reduction in the availability of plant nutrients such as phosphorus, calcium, nitrogen, iron, copper, manganese and zinc. Under saline-alkali conditions there may be actually transitional stages, from high salinity-low alkalinity to low salinity-high alkalinity. Under such conditions, the crops may suffer due to high salinity as well as to un-favourable effects of alkalinity.

Reclamation of Saline and Alkali Soils

Schoonover (1959) in his study of soil problems in India, has listed the following technical requirements for reclamation of saline and alkali soils:

1. Adequate drainage.
2. Availability of sufficient water to meet crop use and also leach the salt below the root zone in the soil.
3. Better than average soil management to include perfect land leveling, good bunding for irrigation and advanced agronomic practices.
4. Protection and reclamation to be taken in large blocks.
5. Irrigation water should be of good quality.

Saline Soil Reclamation and Management

Saline soils in which the soluble salts contain appreciable amounts of calcium and magnesium do not develop into alkali soils by the action of leaching water. The reclamation is comparatively easy in such soils. The main problem is to leach the salts downward below the root zone and out of contact with subsequent irrigation water.

Following Methods may be used for Removal of Salts:

Mechanical Methods

(i) Flooding and leaching down of the soluble salts:

The leaching can be done by first ponding the water on the land and lowering it to stand there for a week. Most of the soluble salts would leach down below the root zone. After a week, standing water (dissolved with soluble salts) is allowed to escape. Such, 2 to 3 treatments are given to reclaim highly saline soils. Sometimes gypsum is also added to flood water when the soluble salts are low in calcium to check development of alkalinity.

(ii) Scrapping of the surface soil:

When the soluble salts accumulate on the soil surface, scrapping helps to remove salts. This is a temporary cure and salinity again develops on such lands.

Cultural Methods (Crop, Soil and Water Management)

(i) Providing proper drainage:

If the soil is not free draining, artificial, drains are opened or tile drains laid underground to help wash out the salts.

(ii) Use of salt free irrigation water:

Salt free good quality of irrigation water should be used.

(iii) Proper use of irrigation water:

It is known that as the amount of water in the soil decreases the concentration of salts in the soil solution increases, thus, moisture should be kept at optimum field capacity.

(iv) Planting or sowing of seeds in the furrow:

The salt concentration even in smaller amounts is most harmful to the germinating seedlings. Water generally evaporates from the highest surface by capillarity and hence, these points have maximum salt concentrations. If the seeds or seedlings are planted inside the furrows, they escape the zone of maximum salt concentrations and thus, can germinate and develop properly during their early growth stage.

(v) Use of Acidic Fertilizer:

In saline soil, acidic nature of fertilizers (e.g., Ammonium sulphate) should be used.

(vi) Use of organic manures:

The organic manures have very high water-holding capacity. When sufficient amount of these manures are added the water-holding capacity of soil increases and as a result the conductivity of the soil solution decreases.

(vii) Ploughing and leveling of the land:

Ploughing and leveling of the land increases the infiltration and percolation rate. Therefore, salts leach down to the lower levels.

(viii) Retardation of water evaporation from soil surface:

Water may be conserved in the soil retarding the water evaporation. Thus, salts may remain in the lower level with the water.

(ix) Growing of salt tolerant crops:

- High salt tolerant crops: Para grass, barley, sugar beet, etc.
- Moderately salt tolerant crops: Wheat, rice, sorghum, maize, flax etc.
- Low salt tolerant crops: Beans, radish, white clover etc.
- Sensitive crops: Tomato, potato, onion, carrot etc.

Reclamation and Management of Alkali (Saline-alkali and non-saline-alkali) Soils

Alkali soils cannot be reclaimed by mere flooding the land. In the case of saline-alkali soils, flooding is likely to do more harm. Leaching (flooding) down of soluble salts make the soil alkaline (only Na-clay remain in the soil). Soils get dispersed and become compact (impervious).

In alkali (non-saline-alkali) soils, exchangeable sodium Na-clay is so great as to make the soil almost impervious to water. But even if water could move downward freely in alkali soils, the water alone would not leach out the excess exchangeable sodium. The sodium-cation must be replaced by calcium-cation and then leached downward.

Following chemical methods are used for reclaiming the alkali soils:

(A) Chemical Methods:

(i) Application of gypsum:

By cationic exchange, calcium is often used to replace sodium in alkali soil. If the soil has no reserve of calcium carbonate, the addition of gypsum (calcium sulphate) is necessary. When gypsum is used as a reclaiming agent, calcium replaces the exchangeable sodium and converts the clay back into calcium-clay (Ca-clay).

$$\begin{matrix} Na^+ \\ \\ Na^+ \end{matrix} \boxed{\begin{matrix} CLAY \\ COMPLEX \end{matrix}}_{\text{Alkali soil}} + \underset{\text{Gypsum}}{CaSO_4} \rightleftharpoons \underset{\text{Normal}}{Ca^{++}} \boxed{\begin{matrix} CLAY \\ COMPLEX \end{matrix}} + \underset{\text{Leachable}}{Na_2SO_4} \downarrow$$

$$Na_2CO_3 + CaSO_4 \rightleftharpoons CaCO_3 + Na_2SO_4 \downarrow$$

Sodium sulphate goes into solution and is then removed by washing it out with water or leaching down with water with the help of artificial drains. Addition of gypsum improves physical conditions of soil. Soils become flocculated and drainage improves. pH is lowered down to a desirable level.

Gypsum requirement is alkaline soil:

For reasonable crop production on a sodic soil, the lowering of the ESP to the level of 10 is considered sufficient. The amount of gypsum required to be added to a sodic soil to lower the ESP to a desired value is known as gypsum requirement. It is expressed in milliequivalent of Ca^{++} per 100 gm. of soil. Gypsum requirement can be calculated from the data on CEC and ESP of the soil.

For a sodic soil, suppose, CEC = 30 and ESP = 60, gypsum requirement to lower the ESP to 10, will be:

$$\left| \frac{60-10}{100} \right| \times 20$$

or = 10 m.e. of Ca^{++} per 100 gm. soil.

Besides gypsum that is best soil amendment for sodic soil, several other materials may be used for reclaiming alkaline soils.

Gypsum equivalents of some such materials are given below:

Amendment	Gypsum equivalent
1. Gypsum ($CaSO1.2H_2O$)	1.00
2. Sulphur (S)	0.19
3. Sulphur acid ($H_2 SO1$)	0.57
4. Iron sulphate (Fe $SO1.7H_2O$)	1.62
5. Iron pyrite(Fe SO_2)	0.63

(ii) Use of sulphur:

In the case of alkali soil that contains free calcium carbonate, addition of sulphur, sulphuric acid, iron and aluminium sulphate, green manure (produce acidity) etc. reclaim the soil very effectively. The acidity developed during the course of their decomposition of soil, neutralizes alkalinity. At the same time brings calcium carbonate into solution which then reacts with the sodium clay and converts it into calcium clay.

When sulphur is spread on the soil, it is oxidised to sulphuric acid, which converts sodium carbonate into sodium sulphate. If calcium carbonate is not present in the soil, it should be added artificially when sulphur is used for reclamation.

Reactions are as follows:

$$\underset{\text{Sulpur}}{2S} + 2H_2O + 3O_2 \xrightarrow[\text{oxidation}]{\text{biological}} \underset{\text{Sulphuric acid}}{2H_2SC}$$

$$\underset{\text{üüüüüü}}{Na_2CO_3} + H_2SO_4 \rightleftharpoons \underset{\text{Leachable}}{CO_2 + H_2O + Na_2SO_4} \downarrow$$

$$\underset{\text{Calcium carbonate}}{2CaCO_3} + H_2SO_4 \rightarrow \underset{\substack{\text{Calcium} \\ \text{Sulphate}}}{CaSO_4} + Ca(HCO_3)_2$$

In mentioned both cases, it is necessary to leach out the sodium salts, formed as a result of bases exchanges with the help of artificial drains.

(iii) Addition of organic matter:

The addition of organic matter increases acidity, thus, helping in lowering the pH. Organic matter is especially helpful where sulphur is added to correct the alkalinity. The organic matter supplies food for the bacteria that stimulates the oxidation of sulphur to the sulphate form. The combination of sulphur, organic matter and gypsum has also been used with success.

(iv) Use of sulphuric acid:

Sulphuric acid changes the sodium carbonate to the less harmful sulphate and also tends to reduce the intense alkalinity. It should be used in the presence of calcium carbonate.

$$Na_2 CO_3 + H_2 SO_4 \rightleftharpoons CO_2 + H_2 + \underset{\text{Leachable}}{Na_2 SO_4} \downarrow$$

(v) Addition of molasses:

Addition of molasses in the soil provide the source of energy for microorganism which on fermentation produce organic acids. The organic acids reduce alkalinity.

(vi) Use of Pyrite:

Pyrite is a mineral containing iron and sulphur and generally it has a chemical composition of FeS_2. Pyrite is found all over the world in igneous and metamorphic rocks and at some places as sedimentary deposits as well.

Pyrite is pyrophoric in nature, produces sulphuric acid and iron sulphate on coming in contact with air and water. The sulphuric acid so produced reacts with the native $CaCO_3$ of these soils to produce soluble calcium which then replaces sodium from the exchange complex.

$$2 FeS_2 + 2 H_2O + 7 O_2 → 2 FeSO_4 + 2 H_2SO_4$$

Pyrite application in non-calcareous alkali soil is not affective because they lack free $CaCO_3$ to be dissolved by H_2SO_4 to produce Ca needed for the replacement of Na from exchangeable complex of sodic soils.

Pyrite application is recommended in the summer season because oxidation of Pyrite is rapid in the temperature range of 25° to 40°C. The activities of microorganism (Thiobacilli) are high in the above temperature range. Low temperature in winter season retards oxidation.

Pyrite should not be applied in the rainy season or in Paddy field. The activity of microorganism (Thiobacilli) decreases at very low at anaerobic (water logged) condition. The activity of microorganism is high in moist soil with good aeration and moderate temperature.

Dose of Gypsum and Sulphur:

On an average for every one milliequivalent of sodium to be replaced, 1.7 tons of gypsum or 3.2 tons of sulphur is required. The amount of gypsum and sulphur required to replace different amount of exchangeable sodium are given in the following table.

Table: Amount of gypsum and sulphur required to replace, indicated amount of exchange sodium.

Exchangeable Sodium m.e. per 100 gm of soil	Gypsum Tons/acre-ft (4,000,000 lbs of soil)	Sulphur Tons/acre-ft (4,000,000 lbs of soil)
1	2	3
1	1.7	0.32
2	3.4	0.61
3	5.2	0.96
4	6.9	1.28
5	8.6	1.60
6	10.3	1.92
7	12.0	2.21
8	13.7	2.56
9	15.5	2.88
10	17.2	3.20

(B) Cultural Method:

Same cultural practices are followed as described in the reclamation of saline soils.

Characteristics of Saline and Alkali Soils

Saline soils contain excessive amounts of neutral salts like chloride and sulphate of sodium, calcium and magnesium, and less than 15 per cent exchangeable sodium. Their pH is less than 8.5, due to the presence of excessive amounts of neutral salts. Saline soils are in excellent physical condition and are permeable to water.

Saline-alkali soils also contain enough neutral salts like chloride and sulphate of sodium and calcium and magnesium to maintain their pH value below 8.5, even when they contain more than 15 per cent exchangeable sodium.

Non-saline alkali soils contain more than 15 per cent exchangeable sodium. Their pH is more than 8.5 because they do not contain any neutral soluble salts. They are sticky and plastic when wet and very hard when dry. Water does not percolate through them.

Formation of Saline and Alkali Soils

Soluble salts originate in the soil from the decomposition of primary minerals. Whenever drainage is restricted and excess amounts of water evaporate from the surface of the land in the arid regions, neutral soluble salts move upward, along with the upward movement of water, and accumulate on the surface as a white crust.

In non-saline alkali soils, soluble salts are washed down by the limited amounts of rainfall in the arid regions when calcium and other ions are replaced from the clay micelle by sodium ions. The clay micelle is then saturated with sodium ions.

They react with the carbonic acid to form sodium carbonate as shown in the following equation:

$$\text{Na}^+ \underset{\text{Na}^+\text{Na}^+\text{Na}^+\text{Na}^+\text{Na}^+}{\overset{\text{Na}^+\text{Na}^+\text{Na}^+\text{Na}^+\text{Na}^+}{\boxed{\text{MICELLE}}}} \text{Na}^+ + \text{H}_2\text{CO}_3 \rightleftharpoons \text{H}^+ \underset{\text{Na}^+\text{Na}^+\text{Na}^+\text{Na}^+\text{Na}^+}{\overset{\text{Na}^+\text{Na}^+\text{Na}^+\text{Na}^+\text{Na}^+}{\boxed{\text{MICELLE}}}} \text{H}^+ + \text{Na}_2\text{CO}_3$$

This sodium carbonate is hydrolyzed to form sodium hydroxide and carbonic acid as shown in the following equation:

$$\text{Na}_2\text{CO}_3 + 2\,\text{H}.\text{CH} \rightarrow 2\,\text{NaOH} + \text{H}_2\text{CO}_3$$

Since carbonic acid is a weak acid and sodium hydroxide is a strong base, the latter completely dissociates to increase the hydroxyl ion concentration of the soil solution and its pH. Therefore sodium carbonate is responsible for the increase in soil pH to more than 9.0 or even more than 10. The surface becomes black colour due to the dissolution of humus in the alkaline medium.

Effects of Soil Salinity and Alkalinity on the Growth of Crops

Saline and Saline-Alkali soils contain excessive quantities of salts, which increase the osmotic pressure of the soil solution. So root hairs absorb less water from saline soils, which contain excessive amounts of chlorides, harmful for crop growth.

Non-saline alkali soils are in a poor physical condition, so roots suffer from poor aeration. They contain borate, bicarbonate and sodium ions even small amounts of borates are toxic to crops. Excessive amounts 01 bicarbonate and sodium ions are also harmful for crops; they make phosphorus and all micronutrients except molybdenum unavailable to crops.

Control of Soil Salinity and Alkalinity

Underground drainage system should be established.

As saline soils are permeable to water heavy irrigation can be used to wash down the soluble salts by adding gypsum to alkali soils, the sodium ions from the clay micelles are gradually replaced by calcium ions, as shown in the following equation:

$$\text{Na}^+ \underset{\text{Na}^+\text{Na}^+\text{Na}^+\text{Na}^+\text{Na}^+}{\overset{\text{Na}^+\text{Na}^+\text{Na}^+\text{Na}^+\text{Na}^+}{\boxed{\text{MICELLE}}}} \underset{\text{Gypsum}}{\text{Na}^+ + \text{CaSO}_4} \rightleftharpoons \underset{\text{Na}^+\text{Na}^+\text{Na}^+\text{Na}^+\text{Na}^+}{\overset{\text{Na}^+\text{Na}^+\text{Na}^+\text{Na}^+\text{Na}^+}{\boxed{\text{MICELLE}}}} \text{Ca}^{++} + \text{Na}_2\text{SO}_4$$

Soluble sodium sulphate is washed down by heavily irrigating the alkali soils, when the above reversible equation proceeds in the forward direction, which means that the sodium ions are gradually replaced from the clay micelle by the calcium ions, which ultimately saturate the clay micelle when soil alkalinity is controlled.

If the alkali soil contains free calcium carbonate, then powdered sulphur of ferrous sulphate may be added to the alkali soil, first producing sulphuric acid and then, calcium sulphate, as shown in the equations.

$$2S + 3O_2 \rightarrow 2SO_3$$

$$\text{Summing up}: \frac{2SO_3 + 2H_2O \rightarrow 2H_2SO_4}{2S + 3O_2 + 2H_2O \rightarrow 2H_2SO_4}$$

$$FeSO_4 + H_2O \rightarrow H_2SO_4 + FeO$$

$$H_2SO_4 + CaCO_3 \rightarrow CaSO_4 + H_2O + CO_2 \uparrow$$

The calcium from this calcium sulphate gradually replace sodium ions from the clay micelle as shown in the following equation:

$$Na^+ \; \underset{Na^+Na^+Na^+Na^+Na^+}{\overset{Na^+Na^+Na^+Na^+Na^+}{\boxed{MICELLE}}} \; Na^+ + CaSO_4 \rightleftharpoons \underset{Na^+Na^+Na^+Na^+Na^+}{\overset{Na^+Na^+Na^+Na^+Na^+}{\boxed{MICELLE}}} \; Ca^{++} + Na_2SO_4$$

Sodium sulphate is washed down by heavily irrigating the field. Ultimately soil alkalinity is controlled.

Chalky Soil

Chalky soil is comprised mostly of calcium carbonate from sediment that has built up over time. It is usually shallow, stony and dries out quickly. This soil is alkaline with pH levels between 7.1 and 10. In areas with large deposits of chalk, well water will be hard water. An easy way to check your soil for chalk is to put a small amount of the soil in question in vinegar, if it froths it is high in calcium carbonate and chalky. Chalky soils can cause nutrient deficiencies in plants. Iron and manganese specifically get locked up in chalky soil. Symptoms of nutrient deficiencies are yellowing leaves and irregular or stunted growth. Chalky soils can be very dry for plants in the summer. Unless you plan to amend the soil, you may have to stick with drought tolerant, alkaline loving plants. Younger, smaller plants also have an easier time establishing in chalky soil than larger, mature plants.

References

- Soil, science: britannica.com, Retrieved April 16, 2019

- Basic-soil-components: extension.org, Retrieved July 19, 2019

- Systems-of-soil-classification, soil-survey: soilmanagementindia.com, Retrieved January 23, 2019

- Different-types-of-clay-soil: hunker.com, Retrieved June 16, 2019

- Properties-clay-soil: sfgate.com, Retrieved May 29, 2019

- Sandy-soil, how-to-crop, farm-basics: farmersweekly.co.za, Retrieved August 21, 2019

- Amending-sandy-soil, soil-fertilizers, garden-how-to: gardeningknowhow.com, Retrieved February 26, 2019

- Sandy-soil-uses-of-sandy-soil, sand, civil-engineering-materials: civiltoday.com, Retrieved May 3, 2019

- Types-of-plants-grown-in-silty-soil: hunker.com, Retrieved March 18, 2019

- Peat-soils: permaculturenews.org, Retrieved March 13, 2019

- Saline-and-alkaline-soil-nature-characteristics-and-reclamation, alkaline-soil: soilmanagementindia.com, Retrieved May 17, 2019

Chapter 2

Processes in the Soil Environment

The process of soil formation regulated by the effects of place, environment, and history is known as pedogenesis. There are various factors that affect soil formation such as organisms, climate and parental materials, relief and time. In order to completely understand the nature of soil, it is necessary to understand the processes related to it. The following chapter elucidates the varied processes and mechanisms associated with this area of study.

Soil Formation

Rocks were slowly broken down to smaller and smaller pieces by physical weathering, and then decomposed by chemical weathering to form the parent material, which may be defined as the weathered material from which the soil was synthesized.

The parent material was further decomposed by the action of atmospheric gases like water vapour, carbon-dioxide, sulphur dioxide, nitric oxide etc. at different temperatures. Some of the primary minerals present in rocks were altered to form some secondary minerals.

Some were completely decomposed and the products of decomposition recombined with each other to form some secondary minerals, which also included clay minerals. These clay minerals or simply the clay remained intimately mixed with primary minerals, like quartz or sand to form some soil.

Some lower plants like algae, muss, lichens (association of algae and fungi is called lichens) etc. began to grow on the bare rocks and on this thin layer of soil, respiring to produce carbon dioxide, which reacted with water to form carbonic acid that decomposed the primary minerals to form clay.

Later on, higher plants began to grow and continued this process. These plants continued to add organic matter to the soil in the form of leaves, roots etc. which decomposed in the soil to form humus that combined with the clay to form the clay-humus complex.

Hence the colour of the surface soil gradually darkened. This process continued till a darker layer of soil of about one foot in thickness developed. This is known as the A horizon. The parent material, which at this stage, was at a considerable distance from the surface, could not be subjected to the direct action of atmospheric agents like heat, moisture and gases etc., so it continued to decomposed at a slower rate forming a relatively compact layer called the B horizon.

Some clay, and sesquioxide (Sum of the oxides of iron and aluminum is called sesquioxide, R_2O_3) were gradually washed down or eluviated from, the lower portion of the A horizon and were washed in or illuviated in the B horizon. So the lower eluviated portion of the A horizon became the E horizon. The E horizon is usually lighter in colour and texture may be and a little acidic in reaction.

In the earlier stages of soil development, soils were dominated by characteristics which they inherited from the parent material. They are dominated by the acquired characters at the later stages of soil development.

For example, soils which were developed from the basic parent material were rich in basic elements and alkaline in reaction during the earlier stages of Soil development. Later on these basic elements were gradually washed down by high rainfall and the soils ultimately become acidic in reaction at the later stages of their development.

We have been that as the parent material was gradually converted to soil, definite layers developed. These layers are called horizon. A soil profile is a vertical section of the earth's crust that includes different layers called horizons, formed due to the action of soil forming factors and also the deeper layers that influences soil formation.

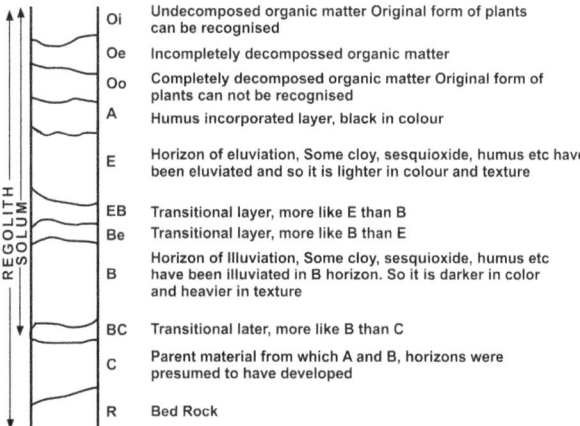

Theoriyical profile

A theoretical soil profile consists of the following horizon:

- O Horizon Organic Horizon develops from dead plant materials on forest lands. This surface horizon occurs just above the mineral or inorganic horizon. It has been subdivided as follows:

- Oi Horizon Consists of under-composed plant material, the original form of which can be recognised.

- Oe Horizon Consists of incompletely decomposed plant material.

- Oa Horizon Consists of completely decomposed plant material, the original form of which cannot be recognised. It is black in colour.

- A Horizon is the top most mineral horizon where humus have been thoroughly mixed and chemically combined with clay to form clay-humus complex. So it is black in colour.

- E Horizon is the horizon of eluviation from which some clay and sesquioxide and humus have been washed down or eluviated. So it is lighter in texture (i.e. respectively richer in sand) and colour.

- EB Horizon Transitional layer between E and B. It is more like E than B.

- BE Horizon Transitional layer between E and B. It is more like B than E.

- B Horizon is the horizon of illuviation. Some clay, sesquioxide humus etc. have been washed in or illuviated in the B horizon, which is relatively heavier in texture and darker in colour. Lime may accumulate in the B horizon in arid region.

- C Horizon is the parent material which is the partially or full weathered material from which A, E and B horizon have been developed.

- R Horizon is the Bed rock.

Solum is the true soil developed by the action of soil forming factors. It includes A, E and B horizon.

Regolith is the unconsolidated material lying about the bed rock. It includes A, E, B and C horizon.

Usually all the above mentioned horizons are not found in every soil profile because they may be immature or over influenced by poor-drainage, topography etc.

Well drained mineral soils usually possess O and Oa if the land is under forest, A or E horizon, B and C horizons depending on soil development.

When farmers being to plough vigin soils, the original layered condition of the soil from 15-25cms depth is destroyed. This layer has been thoroughly mixed up by ploughing and it includes A and E horizon when they are sufficiently deep. Otherwise it may include a little part of B horizon also. This mixed layer is called the furrow slice designated as Ap horizon.

Factors of Soil Formation

Rocks are first broken down to smaller and smaller pieces which are further decomposed to form the parent material. Therefore a parent material may be defined as the unconsolidated, fully or partially weathered material from which the solum has been presumed to have developed.

Parent materials have been classified on the basis of their silica content as acidic or Felsic, intermediate, basic or mafic and ultrabasic or ultramafic Acidic parent material contains more than 66 per cent silica. Intermediate parent material contains 52 per cent to 66 per cent silica.

Basic parent material contains 45 to 52 per cent silica. Ultra basic parent material contains less than 45 per cent silica. Acidic parent material is gradually decomposed by weathering agents to form lighter coloured and textured soils of a relatively poor fertility because the acidic parent material is rich in sand and poor in basic nutrient elements. Basic parent materials are relatively richer in basic elements, plant nutrients and poorer in sand than acidic parent materials.

So they are gradually decomposed by weathering agents to form a darker, heavier (clay) soils of relatively higher fertility and an alkaline reaction. Finer textured soils are richer in humus than coarser textured soils.

The influences of the parent material are more clearly seen in the earlier stage of soil development. But as the time passes basic elements and plant nutrients are gradually washed down from soils that had developed from weathering of basic parent material.

Consequently the soils developed from the basic parent material become poor in basic elements and plant nutrients, and acidic in reaction over a long period.

The nature of soils developed from the weathering of a few important parent materials is as follows.

Lighter Coloured Igneous Rocks

The compositions of Rhyolite, and Granite are almost similar, and therefore, they weather to form acidic parent material, which in turn decomposes to form friable, permeable sandy loam to loam soils of light yellow colour due to their low iron content.

The primary minerals decomposed to form the secondary clay minerals like illite, vermiculite, Montmorillonite etc. in drier climates, which may decompose to kaolinite when both rainfall and temperature increase.

Therefore the soils which have developed in a dry climate may contain fair amounts of plant nutrients and are alkaline in reaction, but those which have developed in a warm humid climate are poor in plant nutrients and moderately to strongly acidic in reaction.

Both trachyte and syenite consists almost entirely of orthoclase. But they also contain low amounts of biotite and hornblende. Since orthoclase weathers very slowly to form clay, the soils developed from weathering of trachyte and syenite is medium in texture and fertility and natural in reaction.

Dark Coloured Igneous Rocks

Andesite, diorite, basalt, gabbro, perdotite and dunite consists of ferromagnesian and calcium bearing minerals, which rapidly weather to form dark coloured clayey soils rich in plant nutrients, basic elements and alkaline in reaction. Usually Montmorillonite and allied mineral smectite (group) dominate the clay minerals. If the rainfall and temperature further increases, the Montmorillonite decomposes to kaolinite.

Sedimentary Rocks

Sandstone mainly consists of quartz, but a little feldspar and mica are also present as impurities.

In general, parent material formed from weathering of sandstone decomposes to form light coloured sandy soils, poor in plant nutrients and acidic in reaction.

Shales are made up of clay minerals and smaller quantities of mica, feldspar, hornblende and quartz. Sometimes a little calcite may be present.

Shales usually weather to from darker coloured, relatively impermeable clayey soils of a high base and plant nutrient status and alkaline in reaction in which usually illite and Montmorillonite dominates the clay minerals. Soils are usually very fertile.

Limestone and dolstone contain more than 50 per cent calcite and dolomite respectively. They weather to form darker coloured clayey soils rich in plant nutrients and basic elements and alkaline in reaction in a cool dry climate.

But the nature of soil formed from the weathering of limestone and dolstone in a warm humid climate depends upon the nature of the impurities present in them because calcium carbonate and calcium magnesium carbonates are washed down by high rainfall and the impurities are left behind.

If the impurities are mainly clay, then dark coloured clayey soils fairly rich in plant nutrients and alkaline in reaction are formed. But if the impurities are quartz or iron oxide, then light coloured, highly permeable sandy soils poor in plant nutrients and basic elements and acidic in reaction are formed.

Metamorphic Rocks

Gneiss usually weathers to form soil which is dominated by smectite and hydrous mica groups of clay minerals in a cool humid climate. These soils contain a fair amount of plant nutrients and are neutral to slightly alkaline in reaction. But these clay minerals further decompose to kaolinite when the temperature and precipitation is increased. Hence highly permeable coarse textured acidic soils are formed in humid climate.

Soils developed from the weathering of mica schists tend to be silty, and are dominated by illite and vermiculite. Hornblende schist may weather to form the smectite group of clay minerals. Schists usually weather to form loamy to clayey soils, fairly rich in plant nutrients and alkaline in reaction.

Recognizable horizons develop faster on sandy loam to loam soils which are high in easily weatherable minerals, are well drained and aerated and are located on moderate slopes. They develop slowly on sandy parent materials which are poor in easily weatherable minerals. They also develop slowly on highly clayey parent materials.

Relation between Parent Material and Vegetation

If the parent material is rich in basic elements, it weathers the soil rich in basic elements and the vegetation growing on it is rich in basic elements. Parent materials like sandstones; quartzites etc. are poor in basic elements Coniferous narrow leaved plants usually grow on such acidic soils in cool humid climate.

Climate and Climatic Water Balance

Weather is the state of the atmosphere at any given time and place. Climate is the average weather. rays striking the earth. Differential heating causes local variation in the earth temperature which cause air mass movement and induces precipitation.

The climate comprises of precipitation and temperature that affect various physical and chemical processes in soil. Climatic water balance is the difference between gain in water mainly through rainfall and loss of water due to evaporation.

This water reached the surface of the earth and took part in soil formation. In arid region, less rain water reaches the ground surface, more water is evaporated and therefore, the climatic water balance decreases.

Climatic Water Balance May be Estimated by Various Means

(i) Lang's Rain Factor $= \dfrac{\text{Mean annual rainfall in millimetres}}{\text{Mean annual temperature in } ^{0}\text{C}}$

(ii) Niederschlag Sathigungs deficit Quotient (NSQ of Meyer (1962)

$\text{NSQ} = \dfrac{\text{Rainfall in Millimetre}}{\text{Absolute saturation deficit of air millimetre of mercury}}$

(iii) Thornthwaite's (1931) Precipitation Effectiveness Index (P.E. Index)

$$\text{P.E. Index} = \sum_{n-1}^{n-12} 115\left(\dfrac{p}{T-10}\right)$$

where p = Monthly rainfall in inches

$\qquad T$ = Mean monthly temperature in degree F.

$\qquad n$ = Cencerned month.

The climatic water balance effect, the decomposition of minerals the formation of clay and the growth of natural vegetation. The soluble products of decomposition are removed along with the percolating and the run-off water.

As the climatic water balance increases, more natural vegetation grows and more organic matter is added to the soil. Hence the organic matter and nitrogen content of soil increase. Primary minerals decompose at increasing rates to form 2: 1 types of clay minerals like smectite (Montmorillonite) hydrous mica and vermiculite groups of clay minerals.

Hence the percentages of clay in soils and their cation exchange capacities increase when the climatic water balance increase.

However, excessive rainfall leaches the basic elements and silica down the soil profile to make soils acidic in reaction when 2:1 types of clay minerals decomposes to form the 1:1 group of clay minerals (Kaolinite) which are further decomposed to hydrous oxides of iron and aluminum, 2:1 types of clay minerals dominated the clay mineralogy of soils of relatively arid regions which are rich in basic elements and alkaline reaction.

A calcium carbonate layer occurs near the surface of the soils if the climate is extremely arid. The depth of this calcium carbonate layer increases with an increase in rainfall. When rainfall is constant and the temperature is increased, less vegetation grows thus less organic matter is added to the soil. Organic matter rapidly decomposes. Hence the humus and nitrogen content of soils decreases.

The increase in temperature also increases the rate of weathering of primary minerals to increase the clay content and cation exchange capacity of young soils. But as the clay decomposes, the cation exchange capacity of old soils decreases.

Living Organisms

The biosphere comprises of billions of microorganisms which improve soil productivity by decomposing rocks, minerals and organic matter. Of special interest in the influence of the biosphere on soil genesis, is the lichens which are symbiotic associations of algae and fungi.

Algae, moss, and lichens grow on bare rock and respire to produce carbon dioxide, which reacts with water to form carbonic acid that dissolves primary minerals and releases the nutrient contained in them for plant growth. Algae use atmospheric nitrogen which is released in the soil upon the death of the algae. Thereafter higher plants grow in an environment richer in nutrients and water.

Higher plants absorb-nutrients from the lower layers of the soil and leave them on the surface in the form of decaying leaves, which decompose, forming humus. Carbon dioxide produced from the decomposition of organic matter and respiration of microorganisms and plant roots continues to decompose primary minerals to form more clay and convert insoluble minerals to soluble ones.

$$CO_2 + H_2O \rightarrow H_2CO_3$$
$$\underset{\text{Insoluble}}{CaCO_3} + H_2CO_3 \rightarrow \underset{\text{Soluble}}{Ca(HCO_3)_2}$$
$$4\underset{\substack{\text{Orthoclase of}\\\text{Microcline}}}{KAISi_3O_3} + 2H_2CO_3 + 2H_2O \rightarrow 2K_2CO_3 + \underset{\text{Keolinite}}{Si_4AI_4O_{10}(OH)_3} + 8SiO_2$$

When plants grow on soils the phosphorus and potassium content of soils increases but the calcium and magnesium content and the soil pH usually decreases.

Root system of grasses are profusely developed and are uniformly distributed throughout the A horizon and the upper portion of the B horizon.

These roots die and decay to form the humus which is incorporated with the soils of the A and the upper portion of the B horizon thoroughly enough to impart dark colour to them. Forest trees shed huge quantities of leaves on the forest floor. Hence forest land soils possess the thick "O" horizon.

Usually grassland soils contain more organic matter and nitrogen and are heavier in texture and darker in colour, of higher cation exchange capacity and pH then the forest and soils.

Pine trees which contain low amount of basic elements, grow on soils poor in basic elements and acidic in reaction.

When both grassland and forest land soils have developed from the same parent material under the same climatic conditions, the rate of removal of bases and the consequent development of soil acidity is more in forest land soil than in the grassland soil probably due to the three following reasons.

Firstly the forest trees return fewer basic elements in the form of decayed leaves to the surface soil. Different species of trees return different amounts of bases to the soil.

Secondly, rainwater leaches the basic elements and plant nutrients to a greater depth in before they are absorbed by roots in forest land soil then in grassland soil.

Thirdly, water entering to rest soils is more acidic in reaction because it contains organic acids formed from decomposition of forest vegetation. The organic acids react with the primary minerals of the forest soil to release the basic ions, which are washed down by high rainfall.

So in forest land soil, the percentage of clay in the B horizon is usually more than that of the A horizon whereas the clay contents of the A and the B horizons of grassland soils are more or less equal.

The decomposition of the organic matter in the forest land soils usually produces organic acids that humus and clay are washed down to B horizon. This horizon therefore possesses illuvial deposits of humus and iron and aluminum compounds in addition to the illuvial deposit of clay.

Soil animals like moles, earthworms, ants and termites bring the soil from B horizon to A horizon. Excessive ploughing of the land, burning of the natural vegetation and overgrazing of the land, decreases the soil productivity and retards soil development.

The application of lime and fertilizers to the soil also retards the development of the soil profile but increases soil productivity. In the humid regions, this practice makes the soil environment more favourable for the growth of grasses then for trees. The soil profile also develops slowly when the soil is constantly mixed up by the animal and the people.

Living organism are influenced both by parent material and climate. Vegetation growing on soils developed from basic parent material is usually acidic parent material. Grasses are the dominant form of vegetation in semi-arid areas whereas the forests are the dominant vegetation in semi humid areas.

Earthworms are usually found in soil richer in basic element and organic matter and alkaline in reaction in the humid region. Ants and termites usually occur in soils of temperate regions and tropical and sub-tropical regions respectively.

Topography

The relief of the land refers to the differences of elevation within it. The relief shown on the topographic map is called the topography of the land. The parent material which is located on the fairly level to very gently sloping lands quickly decomposes to form the soil profile because a major proportion of the rain water washed down some of the soluble salts and humus and fine clay down the soil.

If the parent material has any hard pan to restrict the downward flow of water through it, the Soil profile development will be slower. Water may be ponded on the surface of the level land where the organic matter may accumulate to form the peat and the muck soil.

The rain water rapidly flows down the parent material which is located on the steep slope and carries away considerable amounts of the weathered materials. So less water passes down the parent material on steeper slopes. So less of parent material is decomposed to form less soil on steeper slopes.

When the slope of the land is increased, more clay and basic ions move to the bottom land. Hence the bottom land soils are darker in colour heavier in texture, deeper and more alkaline in reaction than soils on steeper slopes, the soil located on steeper slopes are highly oxidized, red, and acidic in reaction and of shallow depth.

A group of soils which have developed from the same parent material in same climate but under different topographical condition is called Soil Catena. Topography influences vegetation. Grasses occupy the elevated areas and forest plants occupy the lower areas where the climate is neither too dry nor too wet.

Time

Time has been regarded as one of the factors influencing formation soil because even chemical weathering of rocks to form soil requires some time to complete itself.

The parent material passes through the following five stages of soil formation when it gradually decomposes to form soil:

1. Initial stage: The parent material has not yet been weathered.

2. Juvenile stage: Weathering of parent material has just started.

3. Verile stage: Now the easily decomposable parent material has decomposed to form clay. So the percentage of clay in the soil has increased.

4. Senile stage: Slowly decomposable primary minerals have also decomposed.

5. Final stage: Soil profile development is practically complete.

If conditions are favourable, the parent material may be transformed into immature soil or young soil in relatively less time in which organic matter has accumulated in the surface soil and a lesser quantity of parent material has decomposed to form the A horizon; very little of clay sesquioxide and humus has been eluviated from the A horizon.

Only A and C horizons now exist. The soil inherits most of its character from the parent material. As time passes more parent material gradually decomposes to form the B horizon and some clay sesquioxide humus etc. have also been illuviated in the B horizon.

All the horizons have been clearly developed and so the soil is mature. As time passes more primary minerals decompose to form the secondary minerals which are also gradually decomposed to form the hydrous oxides of iron and aluminum. When all the primary minerals have been converted to hydrous oxides of iron and aluminum, the soil has reached the old age stage.

Soil Formation in Arid Regions

Arid regions are commonly characterized by high temperature and low rainfall where chemical weathering of rocks and minerals predominates over their physical weathering. Little decomposition of primary minerals takes place under extremely low rainfall i.e. less than 125 mm per annum to release their constituents in the soluble form that remain in the extremely coarse textured surface soil.

The desert soil thus formed is rich in calcium and magnesium and at some place also in sodium and is therefore alkaline in reaction. A little increase in rainfall i.e. up to 375 mm/annum induces short grasses to come up which get withered during the dry season and rejuvenate with the advent of the short moist season of scanty rainfall.

A little quantity of organic matter in the form of dead roots and shoots of grasses is added to the soil that decomposes to form the humus. The moisture and carbon dioxide generated from the respiration of grass roots and decomposition of dead root facilitates the decomposition of some meager amount of primary minerals to form little amount of clay, Horizons are feebly developed.

Whatever meager amounts of clay and humus have been formed, combine together to impart gray colour to the soil which is known as Sierozems which in Russian mean, gray earth. Its A and B horizons are of gray colour. White specks of lime and gypsum have also been seen in the B horizon. The soil is alkaline in reaction.

Soil Formation in Semi-Arid Region

Temperature is less and annual rainfalls (375 to 1250 mm) are a little more in the semi-arid region than in the arid region. So chemical weathering of primary minerals takes place a little more readily, in the semi-arid region than in the arid region.

Increase in rainfall permits short grasses to come up. They add some organic matter in the form of their dead roots and shoots during the ensuing dry season moisture coupled with the carbon dioxide generated from the respiration of gross roots and decomposition of dead grasses facilitates the decomposition of primary minerals to form some clay.

Humus combines with the clay to impart gray coloured surface soil. This is known as chestnut soil which is rich in calcium and magnesium and is therefore of alkaline in reaction. A lime layer occurs near the surface. A gypsum layer may be present below the lime layer. Hence the soil forming processes is known as calcification.

A little increase in annual rainfall i.e. up to about 1250 mm induces tall grasses to grow densely enough to profusely develop their underground stems and roots. Some roots are regularly sloughed off and decomposed to form humus.

The availability of relatively higher amount of moisture coupled with carbon dioxide generated from the respiration of the underground roots and stems of grass vegetation and also from the decomposition of dead grass roots and stems facilitates the chemical weathering of primary minerals to form the clay.

Humus combines with clay to form the clay humus complex. Hence soils of A and B horizons are uniformly coloured dark gray. The soil is rich in calcium and magnesium and is therefore alkaline in reaction.

Relatively higher rainfall is responsible for lowering the lime and gypsum layer. As the soil is rich in lime, so the soil forming process is known as calcification.

However melanization which is the process of darkening of the soil mass by the in-cooperation of humus with the soil, is also responsible for the genesis of the soil which is the process of darkening of the soil mass by the in-cooperation of humus with the soil, is also responsible for the genesis of the soil which is known as chernozem.

Soil Formation in Humid Climate

Podzolisation is the chemical process of migration of iron and aluminum to the B horizon and precipitation of silica in the A horizon as described below:

The acidic parent material rich in silica and poor in basic elements has developed from the weathering of acidic rocks like quartzite that is rich in silica and poor in basic elements. The weathering of this acidic parent material results in the formation of acidic soils poor in basic elements in cold and moist climates.

Some plants like Hemlock, pine, Heath etc. grow on these soils and deliver organic matter to the soil which is poor in basic elements and therefore decomposes to produce considerable amounts of organic acids, which decomposes clay to liberate silicon, aluminum, iron, and other mediums of the A_2 horizon which becomes ash grey in colour.

Iron, aluminum ions, clay and humus are washed down to the B_2 horizon, where the positively charged iron and aluminum ions react with the negatively charged clay and humic micelle and precipitate there. In Russia, pod means under, zola means ash. They call the process podzolisation, and the soil, podzol because it possesses an ash agey A_2 horizon.

Latosolisation (Laterisatton)

It is the chemical migration of silica from the solum and precipitation of iron and aluminum in the solum as described below.

Basic rocks like Basalt decompose when both rainfall and temperature are high, to form basic parent material, which in turn decomposes to form soil rich in basic the elements, on which some broad leaved forest plants grow and organic matter is added to the soil as dead leaves. The basic parent material is rich in ferromagnesian minerals.

The organic matter is also rich in basic elements. The ferromagnesian minerals and organic matter decompose to release silicon, aluminum, iron and basic ions. The basic ions make the medium alkaline, which keeps the silicon ions in solution which is washed down by high rainfall.

Iron is precipitated and later oxidized to ferric oxide which is red in colour. So the soil may be coloured red. Later, basic elements are gradually washed down when the soils become acidic in reaction, and kaolinite decomposes to hydrous oxides of iron and aluminum.

This process is called latosolisation. Whenever drainage is restricted, a soft deposit of iron oxide occur at or near the water table. It is cut in the form of bricks which harden on drying.

Gleization

When soils remain saturated with water for long periods in the presence of organic matter, ferric compounds like ferric phosphate and ferric sulphide are reduced to ferrous compounds like ferrous phosphate or sulphide which are bluish grey in color.

So this sub-merged soil develops a bluish colour which changes to brown when the soil is re-exposed to the atmosphere. This process of soil formation is known as gleization. Gleization usually takes place in low lying areas where water accumulates.

Salinization, Alkalization and De-alkalization

Soluble neutral salts of sodium, calcium and magnesium originate mainly from the decomposition of primary minerals in the soils of arid regions. When water evaporates from the surface of the soil, the water containing soluble salts moves from the deeper layers to the surface of the soil and deposits them at the surface of the fields, which are covered with a white crust of soluble salts in patches. This process of soil development is known as salinization (or formation of solonchak).

If these soluble salts are removed to the lower layer by a limited amount of rainfall occurring in the arid regions, then calcium and other ions are replaced from the clay and humic micelle by sodium ions.

Consequently the soil becomes sticky and plastic when wet and very hard when dry, black due to the dissolution of humus in the alkaline medium. This process of soil formation is known as alkalization (or formation of solonez).

If the rainfall increases a little, sodium ions are replaced from the clay and humic micelle by hydrogen ions, the silicate clay is decomposed to release silica which is deposited on the soil particles.

Consequently an ash grey colour develops on the soil. The process a soil formation is known as de-alkalization (or formation of Solod).

Soil is a Natural Body

Soil is the collection of natural bodies occupying portions of the earth's surface, possessing properties due the integrated effects of climate and living matter, action upon parent material as conditioned by relief over periods of time.

Soils are natural bodies that exhibit three dimensional sequences of characteristics. First, the properties gradually change downwards from the surface to the bed rock.

The unconsolidated material lying above the bed rock is called the regolith. The upper portion of the regolith is different from the lower portion.

Being nearer the atmosphere, this upper zone has been subjected to the weathering action of wind, water and heat. Plant roots are found in the zone. Organic matter delivered as fallen leaves on the surfaces and dead roots below the surface are decomposed to form humus.

The humus and partially decomposed organic matter is mixed with the soil at the surface layer. Primary minerals are decomposed to form clay. This upper and biochemically weathered portion of the regolith is called the solum.

Scientists have considered the soil to be a natural body possessing both depth and surface area. The properties of soil change vertically because the intensity of weathering of primary minerals decreases from the surface downwards and also because the organic matter is incorporated into surface layer, and decomposed there.

The characteristics of soils also changes in the horizontal direction due to change in topography and parent material e.g. soils on gently sloping upland is well drained and oxidized whereas the soil on the adjacent bottom land may be poorly drained and, consequently in a reduced condition. Soil developed from granite tends to be coarse textured, whereas soil developed from limestone tends to be fine textured.

Scientists have recognized the variation in characteristics of the soil in the horizontal direction. They have set up classification systems in which the soil is considered to be composed of a large number of individual soils, each with its own distinguishing characteristics.

An individual soil body is bounded laterally by other soil bodies. Adjacent soil bodies may be differentiated on the basis of the depth of solum e.g. a soil individual may have depth of 60 cm to 90 cm which means that it is bounded laterally by two other individual soils whose depths are less than 60 cm and more than 90 cm respectively.

All these concepts are applied to find out the minimum size of a soil individual, which is called a Pedon. The lower limit of a pedon is the depth of the solum, its lateral dimensions are large enough to permit the study of the nature of any horizon present because a horizon may be variable in thickness, or may even be discontinuous.

Its area range from 1 to 10 square meters, depending upon the variability in the horizon. Where horizons are intermittent or cyclic and recur at a linear interval of 2 to 7 metres (7 to 23ft), the pedon includes one-half of the cycle. Thus each pedon includes the range of horizon variability that occurs within these small areas.

Where the cycle is 2 metres or all the horizons are continuous and of uniform thickness, the pedon has an area of 1 square metre. Again, under these limits, each pedon includes the ranges of horizon variability associated with that small area. The shape of the pedon is roughly hexagonal. One lateral dimension should not differ appreciably from any other.

Soil as a Medium for Plant Growth

Plant growth depends on the following six factors:

- Light

- Mechanical support

- Nutrient supply

- Water supply

- Oxygen supply an

- Heat

Plant growth depends on the soil for all except the first factor. Roots go deeper in to the soil in search of nutrients and water, Roots anchored in the soil enable growing plants to remains erect.

They cannot grow deep in poorly drained soil and absorb nutrients and water from the soil. Not less than sixteen elements are essential for the growth of crops. They get carbon and oxygen from the air, hydrogen from water and the remaining 13 elements from the soil which are called plant nutrients.

Nitrogen, phosphorus, potassium, calcium, magnesium and sulphur which are required by crops in relatively larger amounts are called major elements or macronutrients and the remaining seven elements i.e. iron, manganese, zinc, copper, boron and molybdenum and chlorine, called minor elements or micronutrients are required in relatively smaller amounts.

About 500 ppm or more of macronutrients are required whereas about 50 ppm or less of micro-nutrients is required by crops. Nitrogen, phosphorus and potassium which are added to the soil in the form of fertilizer, are called fertilizer elements. Calcium and magnesium, are added to the soil in the form of lime are called lime elements.

Primary minerals and organic matter are decomposed to release the nutrients to the soil water in the form of ions. This soil water containing nutrient ions is known as soil solution.

Ionic forms of plant nutrients

	Nutrients	Ionic Form
Crops absorb an enormous quantity of water, most of which is most by transpiration. Only a	Nitrogen	$-NO_3, -NH_4^-$
	Phosphors	$-H_2PO_4 -HPO_4^{--}$
Small fraction of its becomes an integral part of the crop. Therefore the soil must retain sufficient water for absorption by plants.The soil should neither be acidic nor too alkaline for the growth of crops it should be well drained and aerated and free from pathogens.	Potassium	$-K^{++}$
	Calcium	$-Ca^{++}$
	Magnesium	$-Mg^{++}$
	Sulphur	$-SO_4$
	Iron	$-Fe^{++}$
	Manganese	$-Mn^{++}$
	Zinc	$-Zn^{++}$
	Copper	$-Cu^{++}$
	Boron	$-BO_3--$
	Molybdenum	$-MoO_4---$
	Chlorine-CI⁻	

Soil Water

Soil acts as a sponge to take up and retain water. Movement of water into soil is called infiltration, and the downward movement of water within the soil is called percolation, permeability or hydraulic conductivity. Pore space in soil is the conduit that allows water to infiltrate and percolate. It also serves as the storage compartment for water.

Infiltration rates can be near zero for very clayey and compacted soils, or more than 10 inches per hour for sandy and well aggregated soils. Low infiltration rates lead to ponding on nearly level ground and runoff on sloping ground. Organic matter, especially crop residue and decaying roots, promotes aggregation so that larger soil pores develop, allowing water to infiltrate more readily.

Permeability also varies with soil texture and structure. Permeability is generally rated from very rapid to very slow. This is the mechanism by which water reaches the subsoil and rooting zone of plants. It also refers to the movement of water below the root zone. Water that percolates deep in the soil may reach a perched water table or groundwater aquifer. If the percolating water carries chemicals such as nitrates or pesticides, these water reservoirs may become contaminated.

Table: Permeability classification system.

Permeability class	Rate (inches/hour)
Very rapid	Greater than 10
Rapid	5 to 10
Moderately rapid	2.5 to 5
Moderate	0.8 to 2.5
Moderately slow	0.2 to 0.8
Slow	0.05 to 0.2
Very slow	Less than 0.05

Infiltration and permeability describe the manner by which water moves into and through soil. Water held in a soil is described by the term water content. Water content can be quantified on both a gravimetric (g water/g soil) and volumetric (ml water/ml soil) basis. The volumetric expression of water content is used most often. Since 1 gram of water is equal to 1 milliliter of water, we can easily determine the weight of water and immediately know its volume. The following discussion will consider water content on a volumetric basis.

Saturation is the soil water content when all pores are filled with water. The water content in the soil at saturation is equal to the percent porosity. Field capacity is the soil water content after the soil has been saturated and allowed to drain freely for about 24 to 48 hours. Free drainage occurs because of the force of gravity pulling on the water. When water stops draining, we know that the remaining water is held in the soil with a force greater than that of gravity. Permanent wilting point is the soil water content when plants have extracted all the water they can. At the permanent wilting point, a plant will wilt and not recover. Unavailable water is the soil water content that is strongly attached to soil particles and aggregates, and cannot be extracted by plants. This water is held as films coating soil particles. These terms illustrate soil from its wettest condition to its driest condition.

Several terms are used to describe the water held between these different water contents. Gravitational water refers to the amount of water held by the soil between saturation and field capacity. Water holding capacity refers to the amount of water held between field capacity and wilting point. Plant available water is that portion of the water holding capacity that can be absorbed by a plant. As a general rule, plant available water is considered to be 50 percent of the water holding capacity.

The volumetric water content measured is the total amount of water held in a given soil volume at a given time. It includes all water that may be present including gravitational, available and unavailable water.

The relationship between these different physical states of water in soil can be easily illustrated using a sponge. A sponge is just like the soil because it has solid and pore space. Obtain a sponge about 6 x 3 x 1/2 inch in size. Place it under water in a dishpan, and allow it to soak up as much water as possible. At this point, the sponge is at saturation. Now, carefully support the sponge with both hands and lift it out of the water. When the sponge stops draining, it is at field capacity, and the water that has freely drained out is gravitational water. Now, squeeze the sponge until no more water comes out. The sponge is now at permanent wilting point, and the water that was squeezed out of the sponge is the water holding capacity. About half of this water can be considered as plant available water. You may notice that you can still feel water in the sponge. This is the unavailable water.

Water in the form of precipitation or irrigation infiltrates the soil surface. All pores at the soil surface are filled with water before water can begin to move downward. During infiltration, water moves downward from the saturated zone to the unsaturated zone. The interface between these two zones is called the wetting front. When precipitation or irrigation cease, gravitational water will continue to percolate until field capacity is reached. Water first percolates through the large pores between soil particles and aggregates and then into the smaller pores.

Available water is held in soil pores by forces that depend on the size of the pore and the surface tension of water. The closer together soil particles or aggregates are, the smaller the pores and the stronger the force holding water in the soil. Because the water in large pores is held with little force, it drains most readily. Likewise, plants absorb soil water from the larger pores first because it takes less energy to pull water from large pores than from small pores.

Use of soil water estimates on a percentage volume basis does not allow for any practical interpretation. Therefore, water is usually converted from a percentage volume basis to a depth basis of inches of water/foot of soil.

Table: Estimated soil water for three soil textures.

| | ~Inches of water/foot of soil ~ | | |
	Sand	Loam	Silty clay loam
Saturation	5.2	5.8	6.1
Field Capacity	2.1	3.8	4.4
Permanent wilting point	1.1	1.8	2.6
Oven dry	0	0	0
Gravitational	3.1	2	1.7
Water holding capacity	1	2	1.8
Plant available	0.5	1	0.9
Unavailable	1.0	1.8	2.6

The table values are derived from laboratory analysis of soil samples. Some of this information is also published in the Soil Survey. Other techniques have been developed to estimate soil water if laboratory data is not available. Generally, field capacity is considered to be 50 percent of saturation and permanent wilting point is 50 percent of field capacity.

Water holding capacity designates the ability of a soil to hold water. It is useful information for irrigation scheduling, crop selection, groundwater contamination considerations, estimating runoff and determining when plants will become stressed. Water holding capacity varies by soil texture.

Table: Range of water holding capacity for different soil textures.

Textural class	Water holdig capacity, Inches/foot of soil
Coarse sand	0.25-0.75
Fine sand	0.75-1.00
Loamy sand	1.10-1.20
Sandy loma	1.25-1.40
Fine Sandy loma	1.50-2.00
Silt loma	2.00-2.50
Silty clay loma	1.80-2.00
Silty clay	1.50-1.70
clay	1.20-1.50

Medium textured soils (fine sandy loam, silt loam and silty clay loam) have the highest water holding capacity, while coarse soils (sand, loamy sand and sandy loam) have the lowest water holding capacity. Medium textured soils with a blend of silt, clay and sand particles and good aggregation provide a large number of pores that hold water against gravity. Coarse soils are dominated by sand and have very little silt and clay. Because of this, there is little aggregation and few small pores that will hold water against gravity. Fine textured clayey soils have a lot of small pores that hold much water against gravity. Water is held very tightly in the small pores making it difficult for plants to adsorb it.

Since soil texture varies by depth, so does water holding capacity. A soil may have a clayey surface with a silty B horizon and a sandy C horizon. To determine water holding capacity for the soil profile, the depth of each horizon is multiplied by the available water for that soil texture, and then the values for the different horizons are added together. These determinations are shown for two soils.

Table: Calculation of water holding capacity for a soil profile.

Depth from Soil surface	Depth of layer	Soil texture	Water holding Capacity	Available water	Available water
Inches	Feet		In/ft	In/layer	In/5 ft
Soil A					
0-6.0	0.5	Loamy fine sand	1.2	0.6	
6.0-24	1.5	Loamy fine sand	1	1.5	
24-60	3	Fine sand	0.7	2.1	
Total					4.2
Soil B					
0-12.0	1	Silty clay	1.5	1.5	
12.0-30	1.5	Silty clay loma	2	3	
30-60	2.5	Loamy sand	1.1	2.7	
Total					7.2

Water relations are greatly affected by cultural practices, but the effect is largely indirect. For instance, tillage breaks down aggregates, decreasing the number of large pores. This would cause a decrease in infiltration rate and percolation, the water content at field capacity would increase, and gravitational water would decrease. If compaction causes an increase in the number of very small pores, unavailable water may increase, and water holding capacity may decrease. As a result, the amount of plant available water would also decrease.

Importance of Soil Water

Water, an excellent solvent for most of the plant nutrients, is a primary requisite for plant growth. Water serves four functions in plants: it is the major constituent of plant protoplasm (85-95%); it is essential for photosynthesis and conversion of starches to sugars; it is the solvent in which nutrients move into and through plant parts; and it provides plant turgidity, which maintains the proper form and position of plant parts to capture sunlight. In fact, the soil water is a great regulator of physical, chemical and biological activities in the soil.

Plants absorb some water through leaf stomata (openings), but most of the water used by plants is absorbed by the roots from the soil. For optimum water used, it is vital to know how water moves into and through the soil, how the soil stores water, how the plant absorbs it, how nutrients are lost from the soil by percolation, and how to measure soil water content and losses.

Soils also serve as a regulated reservoir for water because it receives precipitation and irrigation water. A representative cultivated loam soil contains approximately 50% solid particles (sand, silt, clay and organic matter), 25% air and the rest 25% water. Only half of this water is available to plants because of the mechanics of water storage in the soil.

Structure of Soil Water

Water can participate in a series of reactions occurring in soils and plants, only because of its structural behaviour. Water is simple compound, its individual molecules containing one oxygen atom and two much smaller hydrogen atoms. The elements are bonded together covalently, each hydrogen or proton sharing its single electron with the oxygen. Instead of the atoms being arranged linearly the hydrogen atoms are attached to the oxygen as a 'V' shaped, arrangement $\left[\begin{smallmatrix} H & H \\ & \diagdown \diagup \\ & O \end{smallmatrix}\right]$ and are separated from each other by angle of only $105\mathring{A}°$.

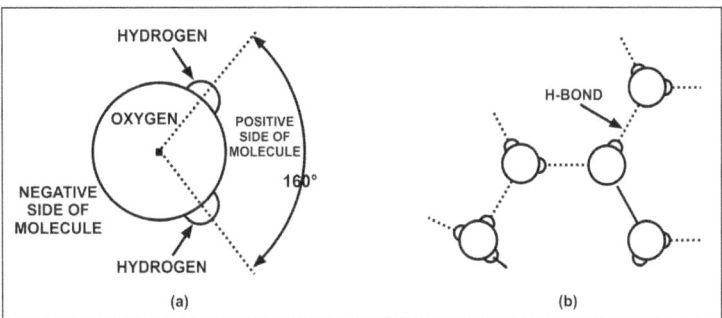

In the above figure (a) the polar water moleculer, H_2O. Because of the non-linear positions of the H's, water is polar. Water has one portion that is more negative than positive and an opposite side that has two hydrogens which are more positive than negative. Polar means there is no centre of zero charge from which there is an equal charge at some distance from that centre in all directions. In (b) the bonding of water to itself through H-bonding is shown, The H of water in soils may bond to oxygen ions of soil mineral surfaces, thereby holding the water tightly to soil. The bonds in (b) become more rigid (less flexible) in colder temperature (ice or snow). A distance between molecules is shown for illustration purposes. In actual shape, molecules are adjacent to each other in close-fit orientation.

Polarity

Due to 'V' shaped structure of water, the side on which the hydrogen atoms are located tends to be electropositive and the opposite side is electronegative. Because of non-linear positions of H+ ions, water is polar. Polar means there is no centre of zero charge from which there is an equal charge at some distance from that centre in all directions.

The H of water in soils may bond to oxygen ions of soil mineral surfaces, thereby holding the water tightly to soil. This accounts for the polarity of water and therefore, water is most important for carrying out many reactions in soils and plants.

Different Forces of Retention of Soil Water

Soil serves as a water reservoir but a leaky one. When too much water is added, the excess runs-off over the surface or into deeper layers. Why does the soil hold some of the water, yet allow part of it do drain deeper? Water is held in soils because of the attraction between unlike charges a positive ion attracted to a negatively charged ion.

The positively charged hydrogen's of water are attracted to nearly negatively charged ions, such as oxygen, even to the oxygen of another adjacent water molecule. Most soil minerals are composed of 70-85% by volume of oxygen. Hydrogen of water bond strongly to these surface oxygen atoms by adhesive bonding (the attraction of unlike molecules).

The hydrogen's of water are also attracted (bonded) to oxygen of other water molecules, including those already adsorbed on to the soil particle surfaces. The attraction of similar or like molecules for each other is cohesive bonding. Such bonding between two molecules through a single hydrogen atom is called hydrogen bonding.

When fatty or oily substances, which are low in oxygen, coat the soil particles, water is not attracted to and held to the coated surface. Such soils are called water-repellent soils. This type of soil are formed in nature under many plant covers and after forest fires, which tend to vaporize oils and resins and drive them into the soil where they coat the soil particles and cause them to resist wetting.

Strong combined adhesion and cohesion forces cause water films of considerable thickness to be held on the surface of soil particles. Because the forces holding water in soil is attractive forces, the more surface (more clay and organic matter) a soil has, the greater is the amount of adsorbed water.

So soil holds water in two ways in the interstices or pores or capillaries between the solid particles, and by adsorption on the solid surfaces of the clay and organic matter.

The mechanism of adsorption of water on the soil surfaces are related to the adhesion and co-hesion forces through hydrogen bonding and also related to the hydration of exchangeable ions which may result in some of them dissociating from the surface into the water.

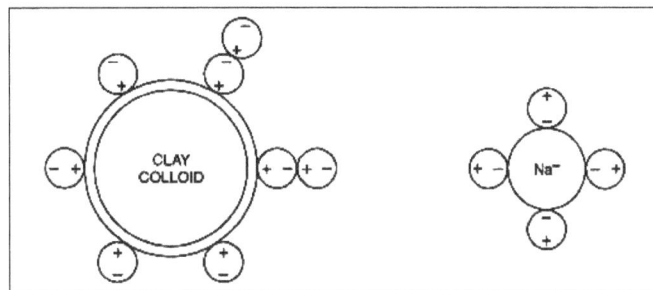

Orientation of water molecules on the surface of clay miclelle and cation

The effect of the cation on the water molecules is greater. The greater its charge and the smaller its size, so the greater its surface charge density, and these effects are influenced by the relative mois-ture content of the clay, by the heat evolved during wetting of clays and by the greater apparent density of the clays in water.

Heat of Solution

When ions are hydrated, a large amount of energy is released and this is known as heat of solution.

Heat of Wetting

When clay particles are hydrated a certain amount of energy must be released and this phenome-non is known as heat of wetting. So there is a close relationship between moisture retention in soil and the energy. The force, with which water is held, is also termed as suction.

Although soil water is held by adsorption and capillary forces, but at the outset it is important to realise that water held in even a fairly dry soil cannot be sharply separated into capillary and ad-sorbed water.

The soil capillaries are not straight uniform tubes, and so for that reason it is better to eliminate the word "capillary" and use the words interstices or pores to describe the spaces between soil particles. So surface tension is an important property, especially as a factor in the phenomenon of capillarity.

The phenomenon of surface tension is generally evidenced at water air interfaces and it may be defined as the forces in dynes acting at right angles to any line of 1 cm length in the surface. At the surface, the attraction of air for the water molecules is much less than that of water molecules for each other.

Consequently there is a net down-ward (in ward) force on the surface molecules, resulting in sort of a compressed film at the surface. This phenomenon is called surface tension.

Consider water in a capillary tube having a boundary with air. The boundary layer between the wa-ter and the air is called meniscus. The meniscus is usually curved; it may make a definite angle, the angle of contact with the walls of the tube; and it puts the water column under a tension T, given by

$$T = 2\ddot{I}f/r \cos \hat{I}\pm$$

If the water is in a circular tube of radius r, where $\ddot{I}f$ is the surface tension of the water and a is the angle of contact or angle of wetting, which is usually zero for the system soil mineral particles-water-air, but may be appreciable if the soil contains much organic matter.

Energy Concepts of Soil Water

The retention and movement of water in soils, its uptake and translocation in plants and potential evapotranspiration etc. are also related to energy. Different kinds of energy art involved including potential, kinetic and electrical. By using the term 'free energy' (ability to do work) energy status of water can be characterized to indicate the strength with which water is held. Several concepts have been used.

The concept of pressure the pressure required to force the water off soil and was measured in atmospheres of pressure needed to remove water. The opposite of pressure- moisture suction or tension. Recently soil water potential is used and it may be defined as the work the water can do when it moves from its present state to a pool of water in the reference state.

The movement of water in soil takes place from a higher free energy to a lower free energy level. It is expected that there is great variability in the free energy levels of water in soils. So the tendency for soil water to move from one soil zone to another due to variation in free energy levels.

Concept of Water Potential

Water in soil has potential energy as well as kinetic energy. Kinetic energy is very small. Potential energy may be defined as the capacity to do work. Pure water has the maximum capacity to do work. Water in soil is held by adsorptive, osmotic and pressure gradient forces and also has relatively lower capacity to do work.

However, work is necessary for the movement of water from one position to another against the force fields to which it is subjected. Extraction of water by plant roots is an example of work done on soil water. Since the term potential refers to the work done per unit quantity, it can be used quantitatively to the work done by water or work done on water as a function of its energy status.

Work is positive when water loses energy and is negative when it gains energy due to movement. For example, a stone sliding down a hill loses potential energy and does positive work, while the stone is moved back up the hill against gravity, it gains potential energy and does negative work.

Adsorbed water is less free to move as compared to water in a pool. Adsorbed water always less free energy (less ability to do work) than water in the pool (zero potential). Therefore, adsorbed water always has a negative potential; work must be done to remove the water to a free pool of water. The more tightly water is adsorbed; the more negative is the number.

Components of Total Soil Water Potential

Based on the concepts of water potential, the total soil water potential can be defined as the work done per unit quantity of pure water in order to transport reversibly and isothermally an infinitesimal quantity of water from a pool of pure water at the reference point to the point under consideration against the force fields.

At equilibrium, the algebraic sum of all forces would be zero. The total soil water potential at any point of equilibrium would be equal to the algebraic sum of all the component potentials as mentioned. Each of the component potentials may be defined in principle, the work done against the respective force field.

The soil water potential is a combination of the effects of the surface area of soil particles and small soil pores that adsorb water, matric potential ($\hat{I}^{..}_m$) the effects of attraction of ions and other solutes for water, solute or osmotic potential ($\hat{I}^{..}_s$) and the atmospheric or gas pressure effects, pressure potential ($\hat{I}^{..}_p$). In salt free well drained soil, matric potential is almost equal to the soil water potential ($\hat{I}^{..}_w$).

An additional effect of the position of the water (such as being elevated) compared to the reference state (the reference free energy state = 0 and is at a specified elevation) is called the gravitational potential ($\hat{I}^{..}_g$). Gravitational potential is not related to soil properties, only to the elevation of water in comparison to a reference position.

Various potentials can be written as follows:

$$\underset{\substack{\text{soil water}\\\text{potential}}}{\Psi_w} = \underset{\substack{\text{matric}\\\text{potential}}}{\Psi_m} + \underset{\substack{\text{solute or osmotic}\\\text{potential}}}{\Psi_s} + \underset{\substack{\text{pressure}\\\text{potential}}}{\Psi_p}$$

$$\underset{\substack{\text{Total water}\\\text{potential}}}{\Psi_t} = \underset{\substack{\text{soil water}\\\text{potential}}}{\Psi_w} + \underset{\substack{\text{soil gravitational}\\\text{potential}}}{\Psi_g}$$

Most of the productive soils have no depth of water standing on it and can be written as follows:

$$\hat{I}^{..}_t â \hat{I}^{..}_w â \hat{I}^{..}_m$$

Therefore, among all potentials matric potential ($\hat{I}^{..}m$) is the most important and dominant for most soils. A relationship between water potential and water content in soil is presented.

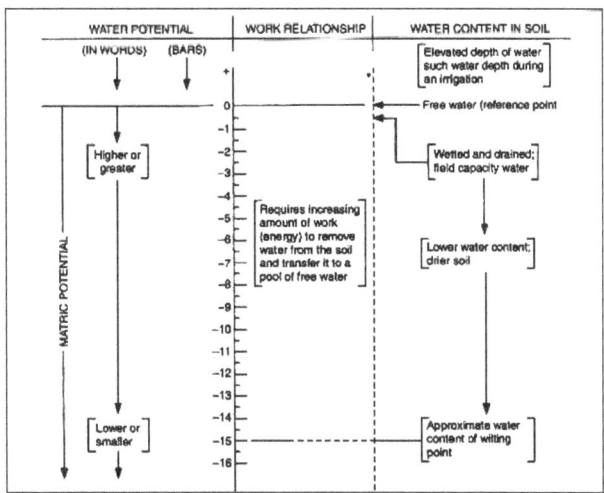

Relationship between water potential and water content in soil. Water with a high potential has more mobility in a soil than has water with lower potential. The matric potential is zero at soil saturation and does riot ever become a positive number

Methods of Expression of Soil Water Potential or Suctions

Soil water potential can be measured in two units at varying energy levels in soil.

(i) pF Scale

The free energy is measured in terms of the height of a column of water required to produce necessary suction or water potential at a particular soil moisture level. So the pF may be defined as the logarithm of centimetre height of a water column to give the necessary suction.

Here 'p' indicates the logarithmic value and 'F' indicates free energy e.g. pF4 is equal to 10,000 centimeters of a water column height (logarithm of 10,000 = 4).

(ii) Bars or Atmospheres

Atmosphere or Bar is the average air pressure at sea level. The term millibar (m bar) is equal to [1/10,000] atmosphere. A popular unit bar is equated to a number of other units as follows:

$\qquad$ 1 bar = 0.9869 atmospheres (approx. 1 atm.)

$\qquad$ = weight of a 1020 cm water column

$\qquad$ = 14.5 lbs. per sq inch

$\qquad$ = 10_6 dynes per sq cm

$\qquad$ = 75.01 cm high mercury column

If the suction is very low as occurs in case of a wet soil containing large amount of water that it can hold, the pressure difference is of the order of about 0.01 atmosphere or 1.01 pF equivalent to 10 cm height of water column. Similarly, if the pressure difference is 0.1 atmosphere the pF will be 2.0.

Table: Relationship among Pf Values, Height of water Column and Pressure or Atmosphere.

Height of water Columns (cm)	Pressure (atmosphere or bars)	pF values
1	0.001	0 saturated soil
10	0.01	1
10^2	0.1	2
346	1/3	2.53 Field capacity
10^3	1.0	3
10^4	10.0	4
15.849	15.0	4.18 Wilting point
31.623	31.0	4.50 Hygroscopic point
10^5	100.0	5
10^6	1000.0	6
10^7	10000.0	7 Oven dry soil

A saturated soil has pF value 0, while an oven dry soil has a pF 7.0.

Classification of Soil Water

There are generally two types of soil water classification based on drying of wet soils and growing plants therein. (A) physical and (B) biological.

Physical Classification

Under physical classification soil water is grouped into three on the basis of retention gravitational, capillary and hygroscopic water.

(i) Gravitational Water

Gravitational water may be defined as the water that is held at a potential greater than -1/3 bar and that portion of the soil water that will drain freely from the soil by the force of gravity. In-spite of having low energy of retention, gravitational water is of little use to plants water occupies the larger pores resulting poor aeration. Therefore, the removal of excess water is a must for the growth of most plants.

(ii) Capillary Water

Capillary water is held in the micro-pores of soils (capillary pores). Capillary water is retained on the soil particle by force of attraction between soil particles and water molecules.

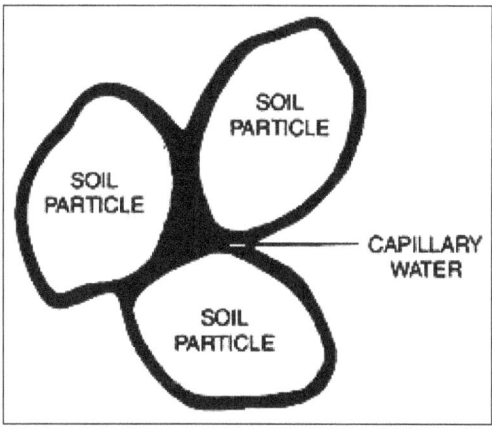

Capillary water held by soil particles

Capillary water is held so rigidly that the force of gravity is not able to separate it from the soil particles. Capillary water is free and moves through the soil pores because of a water potential gradient.

Capillary water may be defined as the water that is retained in the soil between the water potential of 1/3 bar to 31 bars. The force of retention of water molecules by the soil particle is high and part of water is available and part of it is unavailable and so all capillary water is not available to plants.

(iii) Hygroscopic Water

Hygroscopic water is defined as the water that is held by the soil particles at a suction of more than -31 bars. It is essentially non-liquid and moves primarily in the vapour form. This water is held so tenaciously that plants are not able to absorb it and thereby unavailable to plants. Some micro-organisms can utilize such form of water.

Biological Classification

There is a definite relationship between moisture retention and its utilization by plants. Biological classification is based on the availability of soil moisture to the plant. Soil water under this system of classification can be divided into three categories.

(i) Available Water

Available water is defined as that portion of water which is retained in the soil between field capacity (-1/3 bar) and the permanent wilting coefficients (-15 bars). This water is easily usable by plants and therefore, it is called plant available water. Plant available water is equal to the difference of water percentage at field capacity and a permanent wilting point.

(ii) Un-Available Water

Unavailable water is defined as the water which is held at soil water potential greater than -15 bars. It is unavailable to plants. It includes the whole of the hygroscopic water plus a part of the capillary water below the wilting point.

(iii) Superfluous Water

Superfluous water is defined as the water which is retained in the soil beyond the field capacity soil moisture tension. This water includes gravitational water plus a portion of capillary water removed from large interstices. Such type of water is unavailable to plants and rather presence of such water in the soil for a long period causes harmful effect for plant growth because of lack of air.

Soil Moisture Constants

Soil moisture constants and their approximate equivalents in bars of water potential as they affect the relative availability of water of plants.

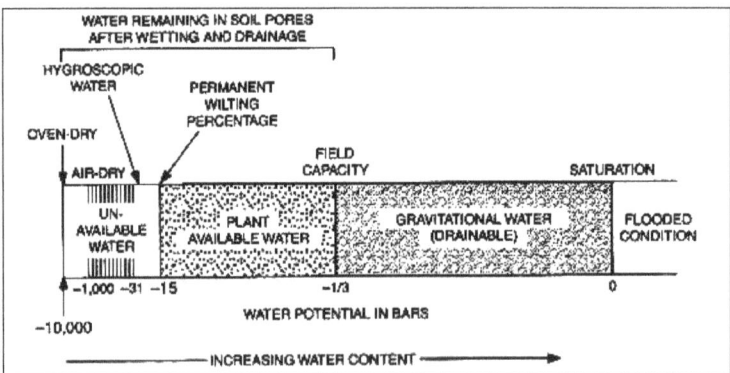

In the above figure soil water constants and their approximate equivalenls in bars of water potential as they affect the relative availability of water to plants. At water potential of -1/3 bars, water is held too loosely to overcome the effect of gravity and drains away. Capillary water (that held by capillary pressure) remains tor plant use; this is held by water potentials ranging from -1/3 bar (field capacity) to -31 bars or lower (hygroscopic), depending upon the pore sizes of the soil_ Plants can only use Capillary water held by not more than -15 bars (the point of permanent witting). Soil water al Vie air-dry Slake is held by water potentials that vary from-1.000 10 -300 bars, depending on humidity.

Oven Dry Weight

Oven dry weight is the basis for all soil moisture calculations. The equilibrium tension of the moisture at oven dryness is 10,000 atmospheres or bars (-10,000 bars of soil moisture potential). It is determined by placing the soil in an oven at 105Â°C until it loses no more water.

Air Dry Weight

Air dry weight is a somewhat variable term, mainly because the moisture in the air fluctuates. Moisture at air dryness is held with a force of 1,000 atmospheres or bars (-1,000 bars of soil-moisture potential). This water is not available to plants.

Hygroscopic Co-efficient

Hygroscopic coefficient is determined by placing an air-dry soil in a nearly saturated atmosphere at 25Â°C until soil absorbs no more water. The soil- moisture tension at this point is equal to 31 bars (soil moisture potential -31 bars) and this water is not available to plants, but available to certain micro-organisms.

Wilting Co-efficient

Sometimes it is also used as permanent wilting point. The wilting point is defined as that amount of water which is held with water potential less than -15 bars and it is held so strongly that plants are not able to absorb it for their needs.

At this point of soil-moisture potential, the plants begin to wilt and at the very beginning of the wilting condition are sometimes recovered with the addition of water and it is then called temporary wilting point, while such wilting condition of the plant is not recovered in-spite of addition of water and then it is called permanent wilting point. Both the wilting points indicate low moisture availability to plants.

Field Capacity

Field capacity is defined as the capacity of a soil to retain moisture against the downward pull of the force of gravity and moisture is held with soil water potential less than -1/3 bar. It is used to determine the amount of irrigation water needed and the amount of reserve soil water available to plants.

Moisture equivalent is approximately equal to the amount of moisture held at field capacity soil. The term moisture equivalent is defined as the percentage of water held by a one centimetre thick moist layer of soil after subjected to a centrifugal force of 1,000 times gravity for half an hour.

Another term "maximum water-holding or maximum water retention capacity" is also used. Maximum water holding capacity is defined as the capacity of a soil to retain water is exceeded. At this point all soil pore spaces (macro and micro pore spaces) are filled up with water and the drainage is restricted.

The water at this point is at a low soil moisture tension. Under natural field conditions only poorly drained soils are at their maximum water holding capacity for long periods of time. Soil saturation, field capacity and wilting points are shown diagrammatically as follows:

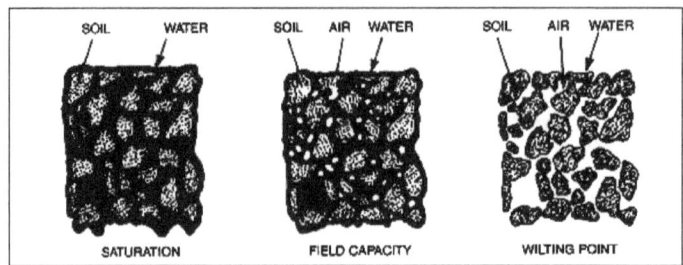

Diagram showing saturation, field capacity and wilting point of soil

Factors Affecting Gravitational, Capillary and Hygroscopic Water

Gravitational

Soil texture and structure are two main factors that affect the amount of gravitational water.

Texture

It plays an important role in regulating the flows of gravitational water. The flow of water is directly proportional to the size of the particles. The larger the size of the particle, the more rapid is the movement of water.

Structure

The different types of soil structure affect the gravitational water by influencing its movement as well as drainage condition of soils. For an example, the rate of movement of gravitational water is slow in platy soil structure which results stagnation of water on the soil surface.

Whereas, spheroidal soil structure helps to improve the movement of gravitational water by increasing its rate of infiltration and percolation. Besides there are other factors like, hard pans in the sub-soil horizon, compactness of soil, organic matter contact in soil etc. also affect the amount and rate of movement of gravitational water.

Capillary Water

There are various factors to be considered that affect the amount of capillary water in soil namely, soil texture, soil structure, surface tension, organic matter content, size of capillary pores in soil, tortuosity (zigzag path) of capillary soil pores etc.

Soil Texture

The finer the texture of a soil the larger the quantity of capillary water it holds. This is mainly attributed to the greater surface area and a large number of micro-pore spaces present in such soil.

Soil Structure

Various types of soil structure present in diversified soils hold water of varying quantities. As for example, soils having platy structure hold excess water as that of granular soil structure.

Surface Tension

An increase in surface tension increases the amount of capillary water. Surface tension is, therefore, an important property and factor that influence the movement and amount of water in the phenomenon of capillarity.

Organic Matter

Organic matter plays an important role for the changes in the capillary water in soil. The presence of organic matter in the soil increases the percentage of pore spaces and consequently increases the capillary capacity of a soil.

Organic matter also influences the soil aggregation as well as formation of soil structure which also affect the amount of capillary water. Humus, a decomposed product of organic matter, has a greater capacity for holding water especially capillary water.

Size of Soil Pores

Different sizes of soil pores hold water with different tenacity. Small and medium sized soil pores tend to hold water with much more tenacity than that of larger size soil pores.

So the movement of capillary water is largely dependent upon the size of capillary pores since different energy levels are associated With Water present in different sizes of pores. Therefore, it affects the availability of such capillary water to the plants.

Tortuosity (zig-zag path) of soil pores and entrapped air in the soil, Soil pores are not continuous, straight and uniform like that capillary glass tubes. Due to such nature of soil capillary pores, the movement of water is somewhat restricted and different. Furthermore, soil pores are field with air which may by entrapped, slowing down or preventing the movement of capillary water.

Hygroscopic Water

The amount of hygroscopic water varies inversely with the size of soil separates. The smaller the size of soil particles the greater the amount of hygroscopic water it adsorbs. Fine texture soils like clay, clay loam soils contain more hygroscopic water as compared to coarse textured sandy soils.

The amount and nature of clay colloids also influence the amount of hygroscopic water. Soils high in colloidal materials (organic and inorganic soil colloids) will hold more hygroscopic water than soils containing low amount of clay and humus. Clay minerals of montmorillonite type having large surface area adsorb more water than that of kaolinite type of clay minerals.

Factors affecting Available Water

The amount of available water is influenced by a number of factors like plant, climatic and soil factors.

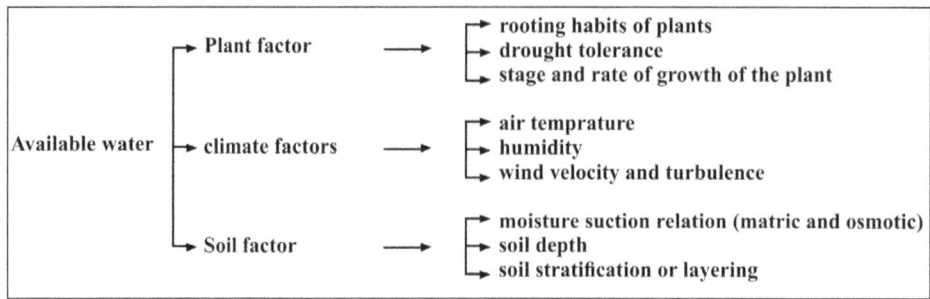

The plant and climatic factors are related to the losses of water vapour under the system known as 'SPAC' (soil-plant-atmosphere continuum). Among the soil factors matric and osmotic suction, soil depth and soil stratification or layering are most important.

Matric Suction

The matric suction means suction due to soil matrix and so the matric suction is influenced by soil texture, structure, organic matter etc. Hence, the texture, structure and organic matter content etc. influence the quantity of available water in soil. The general relationship between soil moisture characteristics and soil texture is shown in the figure.

As the fineness of texture increases, there is a general increase in the amount of available water. The comparative available water holding capacities in relation to water content (inches/foot of soil).

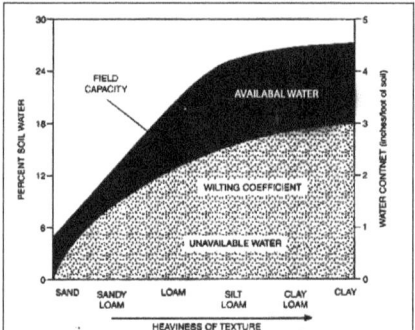

General relationship between soil moisture characteristics and soil texture

Organic matter also influences the amount of available soil moisture storage favourably and this favourable effect is attributable to porosity of soil resulting from well aggregation and formation of good soil structure.

Osmotic Suction

The application of different fertilizers and naturally occurring compounds very often contribute salts to the soil. So the suction develops due to presence of soluble salts in soil and is termed as osmotic suction.

Osmotic suction effects in the soil solution will tend to reduce the range of available water in such soils by increasing the wilting coefficients. The total moisture stress in such soils at this point is matric suction plus the osmotic suction of the soil solution.

Soil Depth

Keeping all other factors equal, deep soils will have greater available water holding capacities as compared to shallow depth soils. Soil stratification or layering will influence significantly the available water and its movement in the soil. Hardpans or impervious layers drastically reduce the rate of movement of water and also influence the penetration of roots adversely.

Hardpans also reduce the soil depth. Sometimes sandy layers also act as barriers to soil moisture movement from the finer textured layers above. Movement through a sandy layer is very sluggish at intermediate and high tensions.

Types of Water Available in the Soil

Gravitational Water or Ground Water

After a heavy rain or on application of irrigation water, the surface layer of a soil is temporarily saturated. This water obeys the laws of gravity and thus descends rapidly through the dry layers, leaving a moist zone in its path.

If there is sufficient water this wet layer penetrates deeply. The rate of this downward percolation of this gravitational water is determined by the number, size and continuity of non-capillary or larger pores.

The gravitational water is available to plants only when rain showers follow one another in rapid succession; otherwise it percolates below the reach of the roots within a few days and thus remains unavailable.

Capillary Water

As ground water drains out of the upper layer of the soil, it leaves behind considerable moisture in the form of films, coating smaller soil particles and as droplets suspended in the angles of larger pores or completely filling the smaller pores.

This water is retained by the forces of surface tension and this capillary water does not respond to the gravitational pull. The forces are, however, small—seldom more than a fraction of an atmosphere depending on the diameter of the capillaries.

This is capillary water. It is primarily this capillary water which is readily available to the plant and this is the source of practically all the water a plant extracts from the soil.

Hygroscopic Water

Owing to evaporation from the soil surface and absorption by roots, the capillary water held by the soil is gradually depleted. As depletion progresses, the forces of molecular attraction or adsorption between the soil particles and the thin films of water held against their surfaces increase until finally the remaining water passes into a state where it is no longer in the liquid condition and thus ceases- to be chemically or biologically active.

The forces of molecular attraction increase rapidly as the water film surrounding soil particle grows thinner until a point is reached when the films are only a few molecules thick and the forces of attraction may reach values as high as 1000 atm.

Evidently plants can absorb only a relatively small amount of this hygroscopic water. The hygroscopic water cannot be entirely evaporated from a soil under ordinary atmospheric conditions, but it can be done by heating soil to a constant weight in an oven at approximately 150°C.

The above three types of soil are not sharply defined but form a continuous series from water which is not retained by the soil, to water which is held with great force.

Chemically Combined Water

The water chemically combined in the structure of soil minerals is known as combined water. After the elimination of hygroscopic water by heating soil to about 150°C., the only water that remains is in the hydrated oxides of aluminium, iron, silicon, etc.

This water is absolutely unavailable to the plants and can only be driven off from the soil by resorting to very high temperature but not before bringing about irreversible changes in the physical and chemical composition of the soil itself.

Types of Movement of Water within Soil

Saturated Flow

Condition of soil, when all large and small pores, are filled with water is called saturated. Saturated flow takes place when the soil is saturated. The direction of flow is from a zone of higher moisture potential to a lower moisture potential.

Generally, water percolates down (vertically) into the lower layers. But horizontal flow also occurs with very less rapidity in comparison to vertical flow. Horizontal movement is much more evident in the clay soil, whereas vertical movement is much more evident in sandy loam.

The movement of percolation water (saturated flow) depends on five factors:

1. Texture,

2. Structure,

3. Temperature,

4. Organic matter, and

5. Pressure.

Texture

The flow of water is proportional to the size of particles. The bigger the particle, the more rapid is the flow or movement. Therefore, water percolates more easily and rapidly in sandy soils than in clay soils.

Structure

In clay soils having single-grain structure (structure-less), the gravitational water percolates more slowly than in those having an aggregate structure (granular or crumby structure). In platy structure, saturated flow is poor in comparison to granular structure.

Organic Matter

Organic matter helps to maintain a high proportion of macro pores. Larger the pore space, greater the flow.

Pressure

The movement of gravitational water is also influenced by the resistance offered by the entrapped soil air. As a result of pockets of air, the soil-air pressure is increased and percolation decreased. This is more especially the case in lower layers. From a practical point of view saturated flow is very important.

Unsaturated Flow

Soil pores contain some air as well as water is called unsaturated soil. Under field conditions most soil-water movement occurs where the soil pores are not completely saturated with water. The soil macro pores are mostly filled with air, and the micro pores with water and some air.

Water movement under these conditions is very slow compared to that occurring when the soil is saturated. Movement will be from a zone of low suction (thick moisture film) to one of high suction (thin moisture film).

The two forces responsible for this suction are the:

1. Attraction of soil solids for water (adhesion) and the

2. Capillarity.

It is, however, observed that the rate of flow decreases with time and at a particular time, becomes very slow, almost negligible. The lowermost portion being more moist than the uppermost one. The force due to gravity is very present.

Therefore, water at any point of the soil column is held by soil particles with a force which is equal to magnitude, by opposite in direction to that of gravity. This force is called Soil suction or soil moisture tension. Water arises from inter-particle pores. Movement is from a zone of low suction to one of high suction or from a zone of thick moisture films to one where the films are thin.

Another explanation of capillary movement is based on free energy concept. When such movement occurs, it does so from an area where the free energy of the soil water is high to one where it is lower. Thus, water movement will occur most easily from soil areas of high moisture level where low attractive forces of the soil, material results in high free energy levels of soil water.

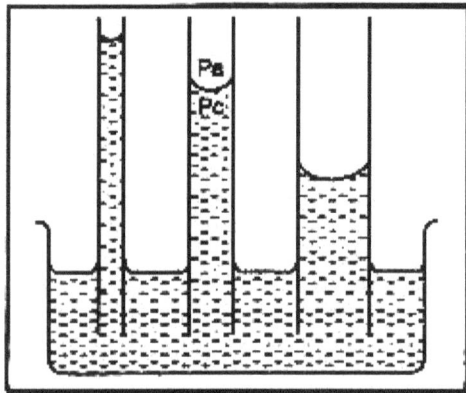

Upward movement by capillarity in glass tubes of different sizes

The movement of moisture takes place because of the difference in water potential. The portion of soil that is moist is at a low water potential, while that which is dry is at a high water potential. Moisture moves from high potential to low potential thus water rises upward in the capillary tube.

Water Vapour Movement

There are two types of water vapour movement:

- Internal movement, the change from the liquid to the vapour state takes place within the soil, that is, in the soil pores, and

- External movement, the phenomenon occurs at the land surface and the resulting vapour is lost to the atmosphere by diffusion and convection (surface evaporation). The diffusion of water vapour from one area to another in soils does occur.

Water vapour moves for high vapour pressure (generally in moist soil) to low pressure (generally in dry soil). Some vapour transfer does occur within soils. The extent of the movement by this means, however, even from one continuous macro pore to another.

References

- Formation-of-soil, formation: soilmanagementindia.com, Retrieved January 10, 2019

- Soil-water-importance-concepts-and-classification, soil-water: soilmanagementindia.com, Retrieved April 15, 2019

- Types-of-water-available-in-the-soil-4-types, types-soil, soil: biologydiscussion.com, Retrieved July 23, 2019

- Main-types-of-movement-of-water-within-soil, soil: soilmanagementindia.com, Retrieved June 26, 2019

Chapter 3

Soil Physics

Soil physics is a domain that deals with the study of physical properties and processes of soil. It also deals with the phases of physical soil components and their dynamics. The study of soil texture, soil structure, organic matter in soil and soil composition fall under this domain. This chapter closely examines these key concepts of soil physics to provide an extensive understanding of the subject.

Soil physics is the study of soil physical properties and processes, including measurement and prediction under natural and managed ecosystems. The science of soil physics deals with the forms, interrelations, and changes in soil components and multiple phases. The typical components are: mineral matter, organic matter, liquid, and air. Three phases are solid, solution and gas, and more than one liquid phase may exist in the case of nonaqueous contamination. Physical edaphology is a science dealing with application of soil physics to agricultural land use. The study of the physical phenomena of soil in relation to atmospheric conditions, plant growth, soil properties and anthropogenic activities is called physical edaphology. Study of soil in relation to plant growth is called edaphology, whereas study of soil's physical properties and processes in relation to plant growth is called physical edaphology. Thus, physical edaphology is a branch of soil physics dealing with plant growth.

The soil physics is a subarea of Agronomy. There are many examples of this specific subject related to Agriculture. This paper will focus, in general, the following cases: (i) erosion, environmental pollution and human health, (ii) plant population and distribution, soil fertility, evapotranspiration and soil water flux density, and (iii) productivity, effective root depth, water deficit and yield.

Erosion, Environmental Pollution and Human Health

Normally, at dry season, from May to August, the air temperature and rainfall are limitations for economical Agriculture under Brazilian politic (no subsides to Agriculture). For these reasons, the cover crop is not common to all farmers. Therefore, the soils present high erodibility.

The exploration of almost all Brazilian annual crops (without irrigation) occurs on wet season, where the main criteria is based on the maximum probability of rainfall to be equal or superior to evapotranspiration at flowering period. Therefore, the majority of sowing dates is done between September and December, where there is tropical precipitation with high intensity (high erosivity) when the leaf area has low value.

The combination of high rainfall erosivity and high soil erodibility is responsible for about 10,000 to 15,000 kg.ha-1.year-1 of erosion on maize, for example. This first millimeter of the soil, in Agriculture, represents chemical products (herbicides and fertilizers, mainly) in the rivers and less soil fertility (less organic matter and nutrients).

The human water consumption in Brazil is around 100 to 500 liters per day per person, where the water caption from the river is common. The environmental pollution caused by erosion prejudices water quality and human health. Therefore, the water and diseases treatments in the cities are necessary.

The alternative crop system and agricultural politic to minimize the soil losses problem is a challenge for the soil physicists (under economical, social and environmental view). The no tillage system could be an option, because the cover crop can protect the rainfall drop impact on the soil, responsible for about 95% (energy balance) of the erosion process. Some changes in the sowing dates and agricultural politic (subsides) must be done to make mulching. It will be benefic for the environment, farmer and the whole society.

Plant Population and Distribution, Evapotranspiration, Soil Fertility and Soil Water Flux Density

The understanding of relationship of plant population and distribution, soil fertility, evapotranspiration and soil water flux density is fundamental to optimize the soil resources. This implies that soil physicists, with agricultural system global vision, could develop techniques to make the crop production system more adequate for each specific environment.

Increasing of plant population demands more water and better plant distribution. Increasing evapotranspiration requires more soil water flux density. The soil fertility depends on the soil volume per plant (plant population and distribution), evapotranspiration (leaf area, species, wind, air temperature and relative humidity, mainly) and soil water flux density.

The maximum water requirement occurs at flowering. The correct plant population is defined as function of the probability of soil water flux density to be equal or superior to maximum evapotranspiration any day in the whole crop cycle.

For high population, the plant distribution becomes more important. The soil physicist must minimize intra specific competition for water and nutrients. The better plant distribution maximizes the soil volume per plant, and the critical content values for all nutrients (soil fertility) are lower. Consequently, the fertilizer requirement decreases.

The corn grain production per plant is constant when there is no intra specific competition for water and nutrients, and the grain production per area has linear increment with the increasing of plant population.

The corn grain production per plant decreases when there is intra specific competition for water and nutrients, and the grain production per area has potential (less than linear) increment with the increasing of plant population. The grain production per plant decrement rate is lower than the plant population increment rate, then the grain production per area increases.

The corn grain production per plant decreases when there is intra specific competition for water, nutrients and light, and the grain production per area decreases with the increasing of plant population. The grain production per plant decrement rate is higher than the plant population increment rate, then the grain production per area decreases.

The point AB shows when the intra specific competition for water and nutrients starts. The correspondent plant population can be larger in better plant distribution, and the maximum grain production per area also can be larger in higher plant population, when the intra specific competition for light starts.

The soil resources (physical and chemical attributes) optimization (plant distribution) is a subject of soil physicists. Soil physicists should start with a dynamic focus and replace the traditional static emphasis. An example is the critical value for potassium (soil fertility).

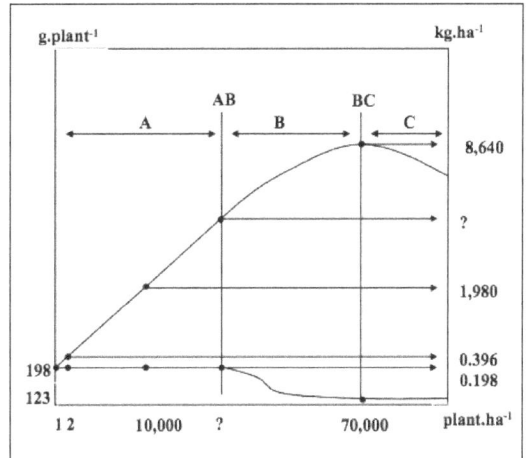

Corn grain production per plant (g.plant⁻¹) and per area (kg.ha⁻¹)
as function of plant population (plant.ha⁻¹)

Plant Population

To better define maize plant population (P, pl.ha⁻¹), the following assumptions were made in conjunction with the optimization:

(i) there is a critical population (Pc, plant.ha⁻¹) where the production per plant (Y, g.pl⁻¹) is constant and production per area has linear increasing (R, kg.ha⁻¹):

$$Y = Ym \ (1 \leq P \leq Pc)$$
$$Ym = Pr_M \ .Fe_M \ .Gf_M \ .Mg_M$$

where Ym (g.plant⁻¹) is the maximum production per plant, Pr_M the maximum prolificity (ear. plant⁻¹), Fe_M the number of grain rows per ear (row.ear⁻¹), Gf_M the maximum number of grains per row (grain.row⁻¹) and Mg_M the maximum grain mass (g.grain⁻¹).

(ii) the production per plant (Y, g.plant⁻¹) and the production per area (R, kg.ha⁻¹), when plant population is larger than critical population (Pc, plant.ha⁻¹), follows the next equations:

$$Y = \frac{Ym}{\left\{1+\left[\alpha\left(P-Pc\right)\right]^{m}\right\}^{n}}, \left(P > Pc\right)$$
$$R = \frac{Y.P}{1000}, \left(P > Pc\right)$$

where α, m and n are the empirical parameters. Therefore:

$$R = \frac{Ym.P}{1000\left\{1+\left[\alpha(P-Pc)\right]^m\right\}^n}, (P > Pc)$$

$$\frac{dR}{dP} = \frac{Ym}{1000}\left\langle \frac{\left\{1+\left[\alpha(P-Pc)\right]^m\right\}^n - Pn\left\{1+\left[\alpha(P-Pc)\right]^m\right\}^{n-1} m\left[\alpha(P-Pc)\right]^{m-1}\alpha}{\left\{1+\left[\alpha(P-Pc)\right]^m\right\}^{2n}} \right\rangle$$

If $\dfrac{dR}{dP} = 0$, then:

$$\left\{1+\left[\alpha(Pm-Pc)\right]^m\right\}^n = m.n.Pm.\alpha^m(Pm-Pc)^{m-1}\left\{1+\left[\alpha(Pm-Pc)\right]^m\right\}^{n-1}$$

$$m.n.Pm.(Pm-Pc)^{m-1} - (Pm-Pc)^m - \frac{1}{\alpha^m} = 0$$

To obtain the solution, the general iterative Newton-Raphson procedure can be used, creating the following function $f(Pm)$:

$$f(Pm) = m.n.Pm.(Pm-Pc)^{m-1} - (Pm-Pc)^m - \frac{1}{\alpha^m}$$

and

$$f'(Pm) = m.(Pm-Pc)^{m-1}\left[n + \frac{Pm.(m=1)}{Pm-Pc} - 1\right]$$

Therefore:

$$Pm_{k+1} = pm_k - \frac{f(Pm_k)}{f'(Pm_k)}$$

To verify the modeled conditions of maize plant population optimization, the second derivation of $R = \dfrac{Ym.P}{1000\left\{1+\left[\alpha(P-Pc)\right]^m\right\}^n}, (P > Pc)$ is given by the following equation:

$$\frac{d\,R}{dP} = \frac{Ym}{1000}\left\langle \frac{g'(P)\cdot\left\{1+\left[\alpha(P-Pc)\right]^m\right\}^{2n} - g(P).2.m.n.a\left\{+\left[\alpha(P-Pc)\right]^m\right\}^{2n\,1}(P-Pc)}{\left\{+\left[\;(\;-Pc)\right]\right\}} \right\rangle$$

where:

$$g(P) = \left\{1+\left[\alpha(P-Pc)\right]^m\right\}^n - m.n.a^m h(P)$$

and

$$g'(P) = m.n.a^m (P - Pc)^{m-1} \left\{ 1 + \left[\alpha (P - Pc) \right]^m \right\}^{n-1} - h'(P)$$

where

$$h(P) = P(P - Pc)^{m-1} \left\{ 1 + \left[\alpha (P - Pc) \right]^m \right\}^{n-1}$$

$$h'(P) = (P - Pc)^{m-1} \left\{ 1 + \left[\alpha (P - Pc) \right]^m \right\}^{n-1} + P.s'(P)$$

$$s(P) = (P - Pc)^{m-1} \left\{ 1 + \left[\alpha (P - Pc) \right]^m \right\}^{n-1}$$

$$s'(P) = (m-1)(P - Pc)^{m-2} \left\{ 1 + \left[\alpha (P - Pc) \right]^m \right\}^{n-1} + (P - Pc)^{m-1} (n-1) \left\{ 1 + \left[\alpha (P - Pc) \right]^m \right\}^{n-2} m \left[\alpha (P - Pc) \right]^n$$

Plant Distribution

Assumptions

To define the better maize plant distribution, the following assumptions were made:

- In the nature, there are only three regular polygons that can stay side by side without empty space: triangle, square and hexagon (a fourth possibility is the rectangle).

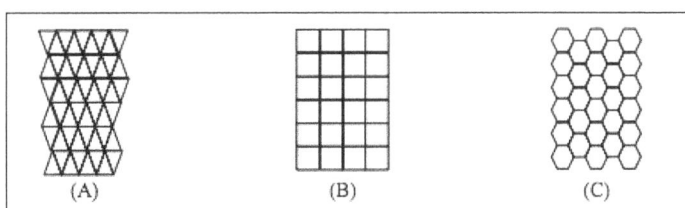

(A) Triangular, (B) square and (C) hexagonal plant distribution

- The maize plant explores circular area

- The better plant distribution, for a fixed plant population, maximizes explored soil area per plant

- Higher soil area per maize plant minimizes stress

- The gross soil area explored by plant (Ap, m².plant⁻¹) is calculated as function of plant population (P, plant.ha⁻¹):

$$AP = \frac{10000}{P}$$

Triangular Plant Distribution

For the triangular distribution, the space between rows (e_1, m) can be calculated as follows:

$$e_1 = \frac{x}{2}$$

By triangle ABC

$$x = 2r\sqrt{3}$$

Substituting $e_1 = \dfrac{x}{2}$ in $x = 2r\sqrt{3}$

$$e_1 = r\sqrt{3}$$

The space between plants (e_2, m) can be calculated as follow:

$$e_2 = 2r$$

The explored useful area per plant (Au, m².pl⁻¹) is calculated as function of the inscribed circle radius r.

$$Au = \pi r^2$$

The gross explored area per plant (Ap, m².plant⁻¹) can be also calculated according to triangle BDE:

$$Ap = \frac{3xr}{2}$$

Substituting $AP = \dfrac{10000}{P}$ and $x = 2r\sqrt{3}$ in $Ap = \dfrac{3xr}{2}$

$$r = \frac{100}{3^{\frac{3}{4}} P^{\frac{1}{2}}}$$

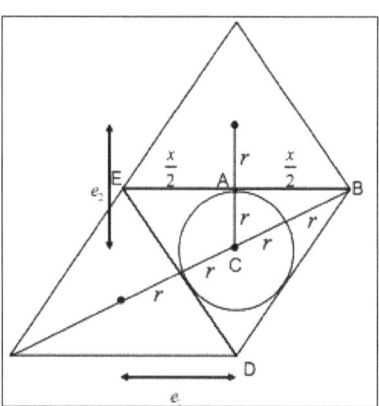

Triangular plant distribution

Substituting $r = \dfrac{100}{3^{\frac{3}{4}} P^{\frac{1}{2}}}$ in $e_1 = r\sqrt{3}$ and $e_2 = 2r$

$$e_1 = \frac{100}{3^{\frac{1}{4}} P^{\frac{1}{2}}}$$

$$e_2 = \frac{200}{3^{\frac{3}{4}} P^{\frac{1}{2}}}$$

Square Plant Distribution

For the square distribution (in figure), the space between rows (e_1, m) and between plants (e_2, m) can be calculated as follows:

$$e_1 = 2r$$
$$e_2 = 2r$$

The explored gross area per plant (Ap, m².plant⁻¹):

$$Ap = x^2$$

Therefore:

$$x = \frac{100}{P^{\frac{1}{2}}}$$

and

$$x = 2r$$

The explored useful area per plant (Au, m²·plant⁻¹) is calculated as function of the inscribed circle radius r:

$$Au = \pi r^2$$

Substituting $x = \dfrac{100}{P^{\frac{1}{2}}}$ and $x \quad r$ in $Au = \pi r^2$:

$$Au = \frac{2500\pi}{P}$$

Substituting $x = \dfrac{100}{P^{\frac{1}{2}}}$ and $x = 2r$ in $e_1 = 2r$ and $e_2 = 2r$:

$$e_1 = \frac{100}{P^{\frac{1}{2}}}$$
$$e_2 = \frac{100}{P^{\frac{1}{2}}}$$

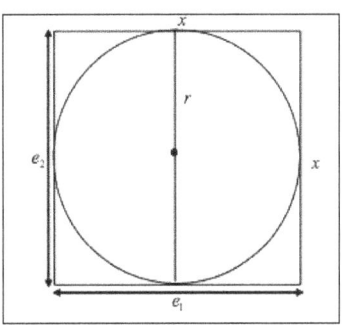

Square plant distribution

Hexagonal Plant Distribution

For the hexagonal distribution, the space between rows (e1, m) can be calculated as follows:

$$e_1 = \frac{3x}{2}$$

By triangle CDE:

$$\alpha = \frac{\pi}{3}$$

By triangle ABC:

$$tg\left(\frac{\pi}{3}\right) = \frac{2r}{x}$$

Therefore:

$$x = \frac{2r}{x}\sqrt{3}$$

Substituting $x = \frac{2r}{x}\sqrt{3}$ in $e_1 = \frac{3x}{2}$:

$$e_1 = r\sqrt{3}$$

The space between plants (e_2, m)

$$e_2 = 2r$$

The gross explored area per plant (Ap, m².plant⁻¹) can be computed as 12 times the triangle ABC area.

$$Ap = 3xr$$

Substituting $x = \frac{2r}{x}\sqrt{3}$ and $Ap = \frac{10000}{P}$ in $Ap = 3xr$:

$$r^2 = \frac{5000}{P\sqrt{3}}$$

or:

$$r = \frac{10}{3^{\frac{1}{4}}}\left(\frac{50}{P}\right)^{\frac{1}{2}}$$

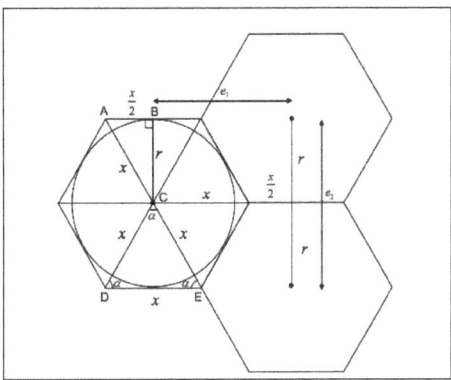

Hexagonal plant distribution

The useful explored area per plant (Au, m².plant⁻¹):

$$Au = \pi r^2$$

Substituting $r^2 = \dfrac{5000}{P\sqrt{3}}$ in $Au = \pi r^2$:

$$Au = \dfrac{5000\pi\sqrt{3}}{3P}$$

Substituting $r = \dfrac{10}{3^{\frac{1}{4}}}\left(\dfrac{50}{P}\right)^{\frac{1}{2}}$ in $e_1 = r\sqrt{3}$ and $e_2 = 2r$:

$$e_1 = 10.3^{\frac{1}{4}}\left(\dfrac{50}{P}\right)^{\frac{1}{2}}$$

$$e_2 = \dfrac{20.}{3^{\frac{1}{4}}}\left(\dfrac{50}{P}\right)^{\frac{1}{2}}$$

Other solutions can be obtained positioning circles minimizing empty spaces. For this particular case the height (h, m) and the area of triangle ABC (At, m²) can be calculated as follows:

$$h = r\sqrt{3}$$
$$At = r^2\sqrt{3}$$

Therefore, there are 2.P triangles ABC per hectare (10.000m²):

$$2.P.r \sqrt{3} \quad 10000$$

Then:

$$r = \dfrac{10.}{3^{\frac{1}{4}}}\left(\dfrac{50}{P}\right)^{\frac{1}{2}}$$

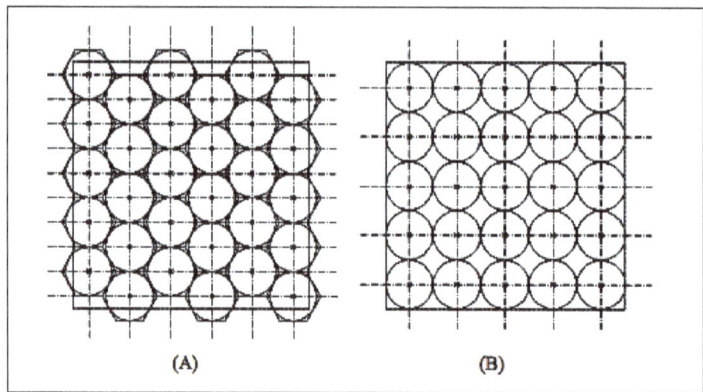

Circles (A) minimizing and (B) maximizing empty spaces

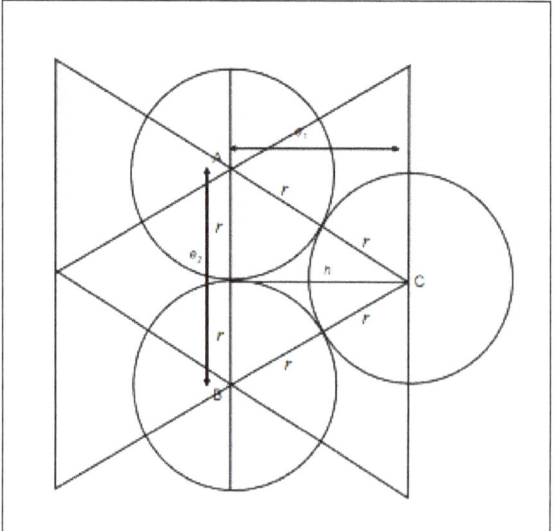

Circles maximizing explored area per plant

Potassium Availability

The potassium availability in the soil could be express in terms of offer rate (kg K.ha^{-1}.day^{-1}) to compare with the crop potassium requirement rate (kg K.ha^{-1}.day^{-1}) (dynamic focus – soil physics contribution) instead critical values for soil potassium content (static emphasis).

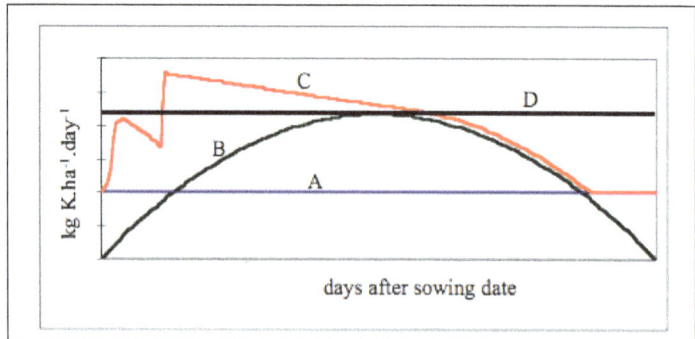

The potassium availability in the soil express in terms of offer rate: (A) deficient soil fertility, (C) deficient soil fertility with two fertilizations, (D) sufficient soil fertility, and (B) the crop potassium requirement rate (kg K.ha^{-1}.day^{-1})

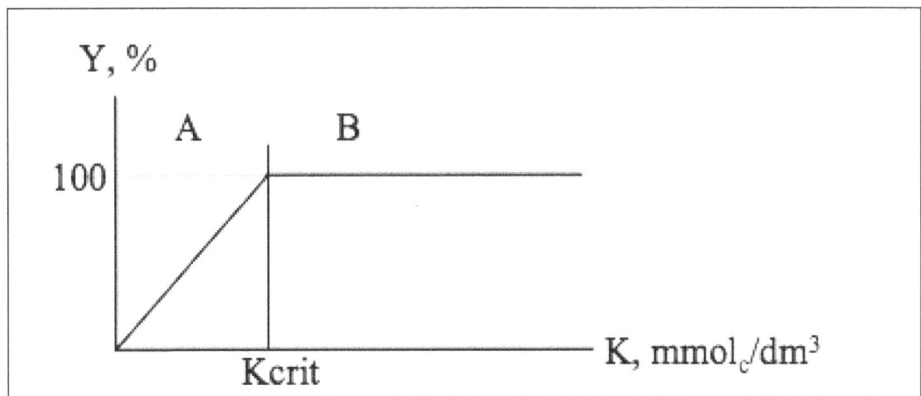

The relative grain yield (Y, %) as function of soil potassium content (K,mmol$_c$.dm³):
(A) deficient soil fertility, (B) sufficient soil fertility, and the critical soil potassium content (Kcrit, mmol$_c$.dm³)

Actually, there is a unique value for the critical soil potassium content (Kcrit, mmol$_c$.dm³) independently of the soil type, specie and weather conditions. This static emphasis was important in the past, but must be replaced per dynamic focus by the soil physicists. It will be an important contribution (it optimizes the fertilizer utilization) for Agriculture.

Productivity, Effective Root Depth, Water Deficit and Yield

Yield means the grain (or other part of the plant) production per area (kg.ha⁻¹), and the productivity will be defined as the potential yield. Then, the productivity depends only of the genotype and weather (soil water content in the field capacity), and yield depends on the genotype, weather and biotic (weeds, diseases and pests, mainly) and abiotic interference.

For practical purposes, the first step is the definition of target yield and price that defines technology level. Then, the first components for agricultural planning at farm scale are: genotype, weather condition (depends on the sowing date), water availability, plant population and nitrogen fertilization.

The water deficit occurs when the soil water flux density is lower than the maximum evapotranspiration. The decreasing of evapotranspiration causes stress. The plant stress reduces yield and increases cost with weeds, diseases and pests control, and decreases profit.

For practical purposes, the soil water holding capacity per unit of effective root depth defines the plant population support with no irrigation agricultural system. When the soil water content is lower than the critical value (θcrit), the soil water flux density (q) is lower than maximum evapotranspiration (ETm) and real evapotranspiration (ETr) decreases.

If the soil water content is larger than θcrit, the soil water flux density (q) is larger than maximum evapotranspiration (ETm) and real evapotranspiration (ETr) is equal to ETm.

The water deficit reduces effective root depth (Ze), because there is no sufficient water to make more roots (the consumption of new cells require more water than old cells). The water excess also reduces Ze, because the oxygen diffusion is limiting (the oxygen diffusion in the air is larger than in the water. The agricultural management must improve effective root depth to optimize natural resources. Each 1 cm soil depth holds around 12,500 L.ha⁻¹ of water.

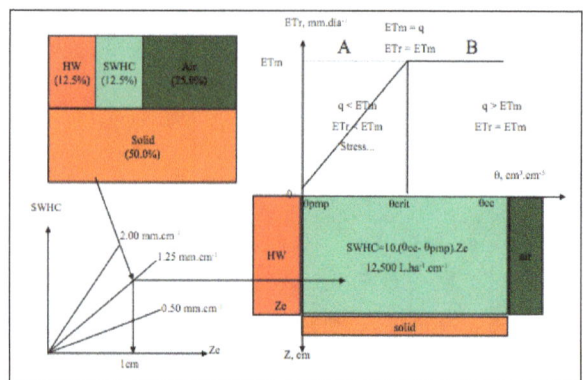

The modal soil physics properties in the nature. HW: hygroscopic water, SWHC: soil water holding capacity, θ: soil water content, ETr: real evapotranspiration, ETm: maximum evapotranspiration, Ze: effective root depth and q: soil water flux density

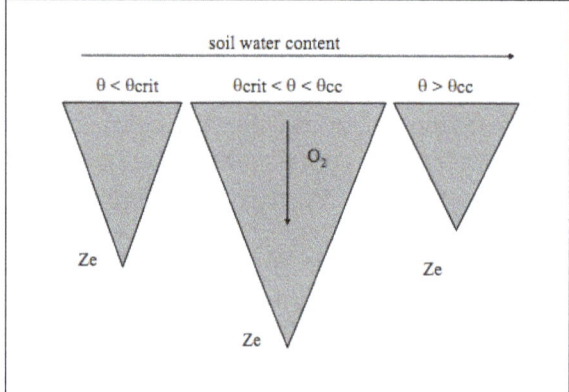

Relationship between oxygen diffusion (O_2), soil water content (θ) and effective root depth (Ze)

Soil Texture

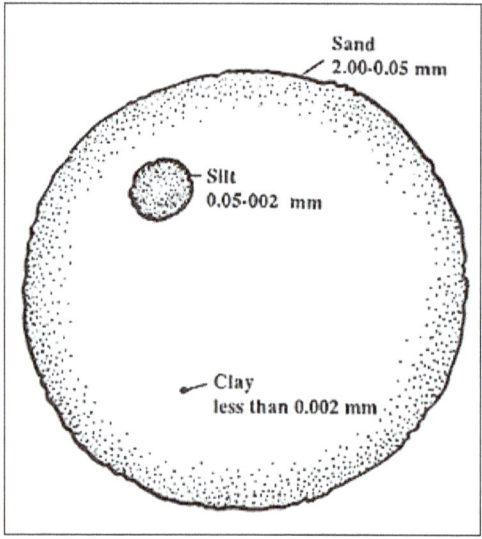

Soil Particle Sizes

Soil texture is about the mineral portion of soil. It takes a thousand years to make an inch of soil. Soil texture is an intrinsic part of soil.

Soil texture is a measurement of the mineral parts of soil, those ground up bits of rock pulverized over the millennia. It is calculated by measuring the proportions of sand, silt and clay particles in soil.

- Sand the gritty bits. Each grain is about the size of the head of a pin.

- Silt smooth and soft like flour. Each particle is about the diameter of a strand of human hair.

- Clay sticky when wet. Each bit is microscopic, about the size of a bacteria.

The sand, silt and clay make up the mineral part of soil. Ideally a soil has about 45% minerals, 25% water, 25% air and 5% organic matter.

Soil texture has an enormous effect on water and on nutrients.

Soil Texture or Soil Biology

Soil experts at the cutting edge say that 90% or more of the function of the soil comes from biology.

Happily organic matter is an effective way to help deal with problems with texture. Organic matter holds nutrients very well. In the water department it can help improve drainage in clay and hold water in sand. It's an all purpose fix.

With compost, mulches and cover crops even the toughest soil textures can come alive.

Soil Texture and Nutrients

This is a gross simplification of one of the ways plants get the nutrients they need. Below is a chart with 17 of the nutrients we know plants need to be healthy.

The non mineral elements, Carbon, Hydrogen and Oxygen, plants get from the air and water. About 95% of the weight of the plant is made up from these three elements. Only about 5% by weight comes from the minerals in the soil. These are still essential to plant growth and health however.

Essential Plant Elements	Symbol	Primary Form
Non Mineral Elements		
Carbon	C	$CO_2(g)$
Hydrogen	H	$H_2O(l)$ H^+
Oxygen	O	H_2O, O_2
Mineral Elements		
Primary Macronntrienis		
Nitrogen	N	NH_4^+, NO_3^-
Phosphorus	P	$HPO_4^{2+}, H_2PO_4^-$
Potassium	K	K^+
Secondary Macronutrielits		
Calcium	Ca	Ca^{2+}

Magnesium	Mg	Mg^{2+}
Sulfur	S	SO_4^{2-}
Micronutrients		
Iron	Fe	Fe^{3+}, Fe^{2+}
Manganese	Mn	Mn^{2+}
Zinc	Zn	Zn^{2+}
Copper	Cu	Cu^{2+}
Boron	B	$B(OH)_3$
Molybdenum	Mo	MoO_4^{2-}
Chlorine	CI	CI^-
Nickel	Ni	Ni^{2+}

Plants take up their nutrients from the soil in ionic form for the most part. If you look closely at the above chart you'll see little + and - signs. The ones with + signs are cations. The ones with - signs are anions. Most of the essential ions are cations or ions with a positive charge.

The sand and silt portions of Soil texture are essentially ground up rock differentiated by their size. These do not have an electric charge.

Clay is different. Clay is a colloidal substance. It is always formed through a chemical weathering process. Each microscopic clay particle carries negative charges along its edges. This is where soil fertility dynamics live.

They act as places to hold cations - the positively charged ions that plants need for their nutrition.

Here is a photo of clay as seen through a scanning electron microscope. It lets us see the flat stacks of clay that create all the great and not so great things about clay.

Uniform vermicular stacks of dickite from Urab

Notice how the particles of clay are flat and stack up on each other. The edges of each of those microscopically thin wafers of clay have negative charges able to attract and hold cations.

Cation Exchange Capacity - CEC

The negatively charged edges of the clay wafers attract the cations, holding them gently in place. By gently we mean kind of like static cling as opposed to say super glue. This keeps the positively charged ions that are plant nutrients in the soil and handy for plant roots to get access to.

The measure of how many cations a particular soil can hold is called the Cation Exchange Capacity or CEC. Having a higher CEC essentially means Soil has a bigger cupboard and can fill the pantry so to speak. In other words it is more fertile.

The amount of clay in Soil determines to a large extent how many nutrients it can hold. The type of clay is also a factor. Clay can vary widely with CEC levels from 5 to 150.

- The Clays - clay, silty clay and sandy clay have a CEC of 15 - 40
- The Clay Loams - silty clay loam, clay loam and sandy clay loam have a CEC range of about 14 -30
- The Silts - silt loam, loam, and silt would range from about 10 - 20
- The Sands - sandy loam, loamy sand and sand would range from 3 - 12

Increase Soil's Ability to Hold Nutrients

Happily soil organic matter, especially humus, has a very high ability to hold nutrients. Its CEC runs 150 to 300. So even though your organic matter is a very small part of the soil it has a significant impact on soil's ability to hold nutrients.

Soil Texture Works with Soil Water

Soil texture has a huge effect on how water moves and is stored or not stored in the soil. There are three types of water dynamic at work here.

Gravitational Water

When it rains or you water your garden water moves into the soil filling all the pore spaces. Over the next few hours or days, depending in part on the soil texture, the excess water will drain away.

Sandy soils have big pores so the water will drain away fast. Clay soils have much smaller pores so gravity drainage may take several days. Plants can use this water while it is present. In sandy soils this happens fast so plants living in sandy soils tend to have roots that can take up the water fast because it drains away fast too.

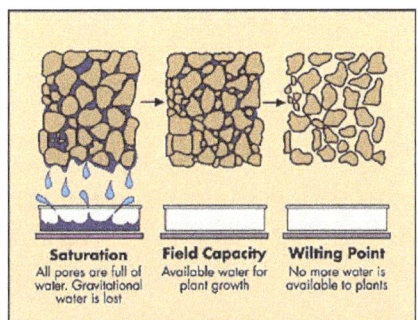

Left is the Gravitational Water Draining Through the Soil, Middle is the Capillary Water Held in the Soil Pores, Right is the Hygroscopic Water Surrounding Each Soil Particle

Capillary Water

Once the excess gravitational water drains away there is still lots of water left in the soil. The water is held in place partly by the tendency of the water molecules to stick to each other and to be attached to the soil particles.

The amount of water in the soil immediately after drainage is done is called the field capacity. Almost all this water is available for plants.

Hygroscopic Water

This is a very thin layer of water surrounding each of the soil particles. This type of water is held so tightly by the soil that it is not available to plants.

A tablespoon full of sand would cover a table while the same volume of clay would cover a football field.

Classes of Soil Texture

Texture names are given to soils based upon the relative proportion of each of the three soil separates sand, silt and clay. Soil that are preponderantly clay, are called clay (textural class), those with high silt content are silt (textural class) those with high sand percentage are sand (textural class). Three broad and fundamental groups of soil texture classes are recognised: sands, loams and clays.

1. Sands

The sand group includes all soils of which the sand separates make up 70 per cent or more of the material by weight. Two specific classes are recognisedâ€"sand and loamy sand.

2. Loams

A loamy soil containing many sub-divisions does not exhibit the dominant physical properties of any of these three soils separates sand, silt and clay. An ideal loam soil may be defined as a mixture of sand, silt and clay particles which exhibits light and heavy properties in about equal proportions.

Note that loam does not contain equal percentages of sand, silt and clay. It does, however, exhibit approximately equal properties of sand, silt and clay.

3.Clay

A clay soil must carry at least 35 per cent of the clay separate and in most cases not less than 40 per cent. For an example, sandy clay soils contain more sand than clay. Similarly silty clay soils contain more silt than that of the clay. Based on these three broad and fundamental groups, the different textural class names developed by U.S. Department of Agriculture and U.S. Bureau of soils.

Table: Textural class names developed by U. S. Department of agriculture

Common name	Texture	Basic soil texture class name
Sandy soils	Coarse	Sandy Loamy sand Sandy loam
Loamy soils	Moderately coarse	Fine sandy loam Very fine sandy loam
	Medium	Loam Silt loam Silt
	Moderately fine	Clay loam Sandy clay loam Silty clay loam
Clayey soils	Fine	Sandy clay Silty clay Clay

Table: Textural Groups on the Basis of Percentages of Sand, Silt and Clay Separates.

Texitiral group	Sand	Silt	Clay
1. Sand	80-100	0-20	0-20
2. Sandy loam	50-80	0-50	0-20
3. Loam	30-50	30-50	0-20
4. Silt loam	0-50	50-100	0-20
5. Sandy clay loam	50-80	0-30	20-30
6. Silty clay loam	0-30	50-80	20-30
7. Clay loam	20-50	20-50	20-30
8. Sandy clay	50-70	0-20	30-50
9. Silty day	0-20	50-70	30-50
10. Clay	0-50	0-50	30-10

Determination of Soil Texture

1. Feel Method

In the field, texture is commonly determined by the sense of feel. The soil is rubbed between thumb and fingers under wet conditions. Sands feel gritty and its particles can be easily seen. The silt when dry feels like flour and talcum powder and is slightly plastic when wet. Clayey particles feel very plastic and exhibit stickiness when wet and are hard under dry conditions.

2. Laboratory Method

A more accurate and fundamental method has been devised by the U.S. Department of Agriculture for the naming of soils based on a mechanical analysis.

Soil Structure

Soil structure refers to the arrangement of soil separates into units called soil aggregates. An aggregate possesses solids and pore space. Aggregates are separated by planes of weakness and are dominated by clay particles. Silt and fine sand particles may also be part of an aggregate. The aggregate acts like a larger silt or sand particle depending upon its size.

The arrangement of soil aggregates into different forms gives a soil its structure. The natural processes that aid in forming aggregates are:

- Wetting and drying,

- Freezing and thawing,

- Microbial activity that aids in the decay of organic matter,

- Activity of roots and soil animals, and

- Adsorbed cations.

The wetting/drying and freezing/thawing action as well as root or animal activity push particles back and forth to form aggregates. Decaying plant residues and microbial byproducts coat soil particles and bind particles into aggregates. Adsorbed cations help form aggregates whenever a cation is bonded to two or more particles.

Aggregates are described by their shape, size and stability. Aggregate types are used most frequently when discussing structure:

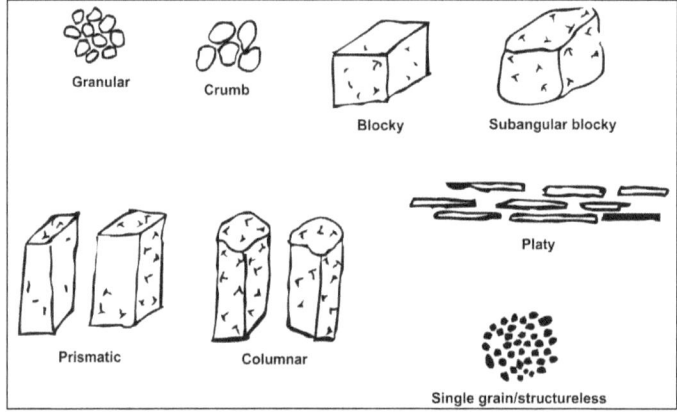

Soil structural types

Table: Structure type and description.

Type	Description
Granular	Rounded surface
Crumb	Rounded surface but larger than granular
Subangular blocky	Cube-like with flattened surfaces and rouned corners

Blocky	Cube-like with flattened surfaces and shape corners
Prismatic	Rectangular with a long vertical dimension and flattened top
Columnar	Rectangular with a long vertical dimension and rouned top
Platy	Rectangular with a long horizontal dimension
Single grain	No aggregation of coarse particles when dry
Structureless	No aggregation of fine particles when dry

Structure is one of the defining characteristics of a soil horizon. A soil exhibits only one structure per soil horizon, but different horizons within a soil may exhibit different structures. All of the soil-forming factors, especially climate, influence the type of structure that develops at each depth. Granular and crumb structure are usually located at the soil surface in the A horizon. The subsoil, predominantly the B horizon, has subangular blocky, blocky, columnar or prismatic structure. Platy structure can be found in the surface or subsoil while single grain and structureless structure are most often associated with the C horizon.

Aggregates are important in a soil because they influence bulk density, porosity and pore size. Pores within an aggregate are quite small as compared to the pores between aggregates and between single soil particles. This balance of large and small pores provides for good soil aeration, permeability and water-holding capacity.

Tillage, falling raindrops and compaction are primarily responsible for destroying aggregates. As the cutting edge of a tillage implement is pulled through the soil, the shearing action at the point of contact breaks apart aggregates. If tillage is conducted at the same depth for several years, a tillage pan may develop. This is one form of compaction. Particles that were once part of the aggregates may reorient themselves and form platy structures. The amount of aggregate destruction that results from tillage depends on the amount of energy the tillage implement places in the soil. The field cultivator has little down pressure and destroys few aggregates. The disk, however, has both cutting action because of the rotation of the disk and shearing action. Together there is substantial down pressure and destruction.

Aggregates on the soil surface can be broken down by the beating action of raindrops. The single particles that were once part of the aggregate can easily form a crust when the soil dries. The crust looks very similar to the crust formed on a puddle after it rains. It is very difficult for water to infiltrate a crust and for seedlings to push up through a crust. Thus, field operations that lead to aggregate destruction at the soil surface have detrimental secondary effects. The particles also can be eroded if they become detached by rainfall.

Compaction can lead to the breakdown of aggregates in the surface soil and subsoil if the applied force from wheel traffic, animal traffic or human traffic is greater than the force holding an aggregate together. Field observations have shown that compaction can cause granular structure on the soil surface to break down and reform as blocky structure and blocky or subangular blocky structure in the subsoil to become structureless.

Aggregation is promoted by root growth and the addition of organic material. Roots excrete compounds that are used as food by microorganisms. Also, as roots absorb water and dry the soil, cracks form along planes of weakness. Lastly, when roots decay, root channels serve as conduits for water that facilitate wetting/drying and freezing/thawing.

Organic material may be added in the form of crop residue, animal manure, sludge, and green manure. These additions are usually made to the surface soil and are critical to the development of granular and crumb structure. As organic material is incorporated by tillage, soil animals and microorganisms, it aids in subsoil structure development.

Soil Composition

The soil composition is one of the most important factors affecting soil properties, nutrient regime of soils and selection system of fertilizers.

For soils we distinguish the following:

- Phase composition (solid, liquid, gas phase),
- Grain composition,
- Chemical composition.

The phase, grain and chemical composition differ with various soils and with its complex effect it determines the formation of various physical, chemical, physicochemical and biological properties of the soil. Especially, it effects:

1. sorption of nutrients in the soil,
2. soil reaction,
3. buffering capacity of soil,
4. soil concentration in soil solution,
5. air and water content in soil,
6. soil temperature,
7. reduction and oxidation processes in soil,
8. soil biological activity,
9. content of available nutrients (macroelements, macroelements, foreign elements).

Solid Phase of Soil

The solid phase of soil consists of a set of solid soil particles of the most varied composition and sizes. It consists of the mineral component which accounts for most of our soils by 95-98% of weight of the dry matter of all the solid soil particles solids. Considerably smaller proportion (2-5%) is the organic portion of soil.

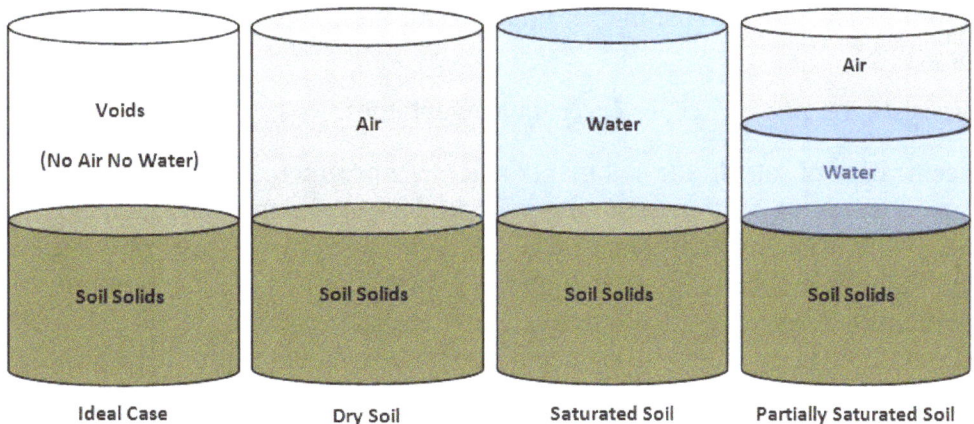

| Ideal Case | Dry Soil | Saturated Soil | Partially Saturated Soil |

Fluid Phase of Soil

The fluid phase of soil poses the soil water which determines the existence of a cycle of substances as an irreplaceable factor for edaphon and vegetation. In terms of nutrition not only the transport of substances from the soil solution into the living cells of the root system of plants is important, but also the vertical thransport through the soil profile. This is the reason for the loss of nutrients by washing away into the lower layers. Not only the nutrients supplied by fertilizers are washed out, but also together with the nutrients of soil supply and those released by the mineralization of soil organic matter and soil-forming processes of mineral mother substrate.

The composition and concentration of the soil solution are the result of a series of physical, chemical, physico-chemical and biological processes in the soil in close relation to temperature, humidity, soil aeration and the composition of the solid phase of soil.

The soil solution depending on the above conditions contains numerous dissolved mineral and organic substances in varying amounts and ratios. From mineral substances these are especially cations such as K^+, Na^+, NH_4^+, H^+, Ca^{2+}, Fe^{2+}, Fe^{3+}, etc., out of which a part can be bond in chelate bonds. The anion component of the soil solution comprises mainly of HCO_3^-, SO_4^{2-}, NO_3^-, $H_2PO_4^-$, OH^-, Cl^-, and very small quantities of certain compounds of molybdenum, boron, etc. Single ions get into the soil solution by the weathering of minerals, ecomposition of organic substances, exchange from sorption complex and from air pollution.

The total amount of salts in the soil solution can fluctuate between several hundredths of a percent to several percent (soils full of salt). In the "healthy soils" the salts make up of about 0,05 % in the soil solution. The composition and concentration of salts in the soil solution varies during the year. An increased sal concentration is mainly caused by fertilization, soil drying and weathering, and an intensive mineralization by organic substances. The reduction of salt concentration is a result of an increase of soil humidity, the outflow of nutrients by plants and microorganisms, their flooding or immobilization into insoluble forms, etc.

The fruits can be classified according to the resistance to salt concentration as follows:

1. With good resistance against higher concentration of salts: beets, rape, cabbage (fodder), sorghum,

2. Medium resistance: cereals, maize, millet, sunflower, asparagus, lettuce, spinach, tomatoes, pepper,

3. Low resistance: peas, vetch, celery, cabbage, potatoes.

At high concentration of salts in the soil plants react by a protracted growth and the formation of small dark green leaves. In relation to the above-ground parts of plants the proportion of root mass is increasing. When over-fertilized, plants receive more water, hence, they rarely show sighns of wilting. Plant damage is dependent on the kind of a salt. All kinds of ions in excess are not toxic the same for plants and significantly different are aslo their interference relationships. The risks of plant damage are limited by injecting fertilizers into the whole tillage profile by plowing. (potassium, phosphorous, calcium, magnesium fertilizers). If not possible, we ensure at least a spacing of 2 – 3 weeks between the fertilization and sowing.

Gas Phase of Soil

The gas phase of soil consists of soil air. It closes the pores without water and compared to the atmospheric air it usually contains more CO_2, less O_2 and increased amounts of water vapour. Although there is a continuous exchange of gaseous components between the soil and the atmosphere, it lacks a continuous balancing of differences.

Carbon dioxide in the soil air reaches an average of 0.3%. Insufficient aeration conditions may result in CO_2 content of 1-5%. The CO_2 content in the soil increases due to the decomposition of organic substances by soil microorganisms, roots breathing, and insufficient aeration. Carbon dioxide with water forms carbonic acid (H_2CO_3), which is an important regulator of soil reaction (pH range of 5.2 to 6.5) and also acts directly during the plant nutrition by carbon.

The oxygen content in the soil air is in the range of 10 to 20% and ensures breathing of all soil organisms and is used for the oxidation of organic and mineral substances. The lack of oxygen especially leads to a reduction of ferruginous and manganese compounds. The reduction is supported by the presence of organic residues in poor aeration and also affects other compounds (SO_4^{2-}).

The nitrogen content in the form of NH_3 is also increased in the soil air compared with the atmospheric one. Elemental nitrogen (N_2) is used during free and symbiotic fixation. In anaerobic conditions, nitrogen is again released in the form of nitrogen oxides (NO_x) from NO_3^- or as an elemental N_2 denitrification.

Soil Testing

The purpose of soil testing in high-yield farming is to determine the relative ability of a soil to supply crop nutrients during a particular growing season, to determine lime needs, and for diagnosing problems such as excessive salinity or alkalinity. Soil testing is also used to guide nutrient management decisions related to manure and sludge application with the objective of maximizing economic/agronomic benefits while minimizing the potential for negative impacts on water quality.

Sampling

The soil testing program starts with the collection of a soil sample from a field. The first basic principle of soil testing is that a field can be sampled in such a way that chemical analysis of the soil sample will accurately reflect the field's true nutrient status. This does not mean that all of the samples must, or will, show the same test results, but rather that the results must reflect true variations within the field. Remember that the soil test recommendations for lime and fertilizer can never be more accurate than the accuracy of soil sampling.

Factors Affecting Nutrient Availability	N	P	K	S	CA and MG	Micros
Soil pH	X	X	X	X	X	X
Moisture	X	X	X	X	X	X
Temperature	X	X	X	X	X	X
Aeration	X	X	X	X	X	X
Soil Organic Matter	X	X		X	X	X
Amount of Clay	X	X	X	X	X	X
Type of Clay		X	X		X	X
Crop Residues	X	X	X	X	X	X
Soil Compaction		X	X			
Nutrient Status of Soil		X	X		X	
Other Nutrients		X	X		X	X
Crop Type	X	X		X		X
Cation Exchange Capacity (CEC)			X		X	X
% CEC Saturation					X	

Nutrients are only taken up by roots when dissolved in water. Insoluble nutrients are not immediately useful for plant nutrition.

Extraction and Chemical Analysis

Once the soil samples have been collected and prepared, the level of available nutrients in each sample must be determined. Many chemical methods have been suggested and are being used for the measurement of essential plant nutrients. The criteria for chemical extracting and analysis of plant nutrients are that those processes must be rapid, accurate and reliable.

Most chemical extracting methods allow the extracting solution, which may consist of water, alkali, weak or strong acid or combinations of these chemicals, to react with the soil sample in a relatively short time. The sample is then filtered and the solution analyzed for the available nutrients.

Soil Test Parameters

In addition to extracting solutions, several other parameters of each soil test are important in determining the final number that is printed on a soil report for any one soil sample. These parameters include:

- Ratio of soil to extractant.

- Shaking time, action and speed.

- Method of expressing the results (e.g., lb/acre, ppm, index systems).

- "Cut-off" levels for high test results.

- Overall techniques used in the lab.

The extractants containing the dissolved plant nutrients are analyzed to determine the concentration of the plant nutrient(s). Results are usually reported as parts per million (ppm), or pounds per acre (lb/acre). For most nutrients, ppm may be converted to lb/acre by multiplying by two (40 ppm of potassium = 80 lb/acre). For nitrate, sulfate and chloride, essentially all the nutrient forms present in the soil are extracted, and depth increments, other than the standard 6- to 7-inch surface layer, are sampled. For these measurements, ppm is converted to lb/acre by the following formula: lb/acre = ppm x 0.3 x depth increment in inches. For example, a 10 ppm nitrate N test on a soil sample taken to a 24 inch depth would convert to 72 lb/acre (10 ppm x 0.3 x 24 inches). In this case, 72 lb/acre of nitrate nitrogen were present in the top 24 inches of the soil sampled. Extracting available plant nutrients helps give an educated estimate as to the amounts of plant nutrients that will be available to a particular crop during the growing season. The amount of plant nutrients extracted will depend on the strength of the extracting solution and various other parameters. Soil test values are a relative number and should be interpreted as low, medium or high for a particular nutrient.

Calibration and Interpretation

Perhaps the greatest challenge in soil testing is calibration of the tests. It is essential that the results of soil tests be calibrated against crop responses from applications of the plant nutrients in question. This information is obtained from field and greenhouse fertility experiments conducted over a wide range of soils. Yield responses from rates of applied nutrients can then be related to the quantity of available nutrients in the soil.

The results of long-term soil test calibration studies on different soil types are then utilized to establish recommended amounts of plant nutrients to apply to a particular crop at a given soil test level. For instance, if the soil test P level is in the range of 0–10 ppm (which is low), the P recommendation for a 150 bu/acre corn crop may be 100 lb/acre of P2O5; whereas, if the soil test P level is above 40 ppm (very high), the recommendation may be 0 to 20 lb/acre.

In this example to the right, more than 85 percent of the fields testing very low in a particular plant nutrient may give a profitable yield response to the added nutrient. At the very high level, there is only a 15 percent probability of a profitable yield increase to the added nutrient. These values are arbitrary, but they illustrate the idea of expectation of response.

The tools of site-specific precision management now allow growers to manage more homogeneous areas within fields. Some of those areas have much higher yield potentials than the database with which most of today's soil tests were calibrated. This lack of calibration for high-yielding areas is one of the factors driving interest in using yield monitors and global positioning satellites to conduct strip trials to determine the adequacy of existing soil fertility programs. New precision ag tools have the ability to develop algorithms that allow for management of multiple site-specific zones within individual fields. This means a balanced crop nutrition prescription can be delivered to each square foot of every field.

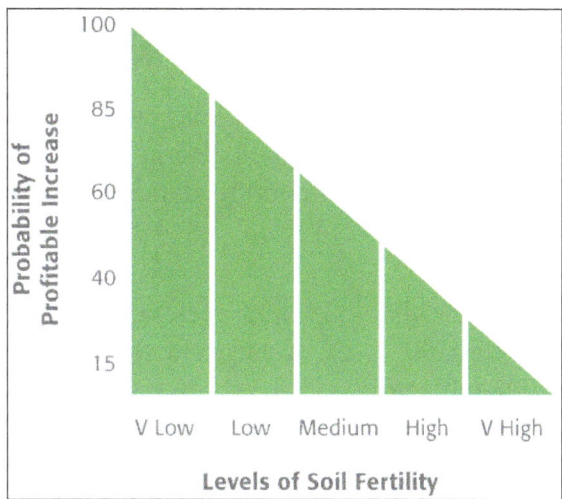

When interpreting soil test results, several things should be kept in mind:

- The chances of getting a profitable response to fertilization are much greater on a soil that tests low in a given nutrient than on one that tests high.

- This does not rule out the possibility of a profitable response from nutrient application at a high level of fertility or lack of a profitable response on soils of low fertility.

- Soil tests are better at predicting the probability of a profitable response to nutrient application than predicting the actual quantity of nutrient that will be needed in any one year.

- Research in the United States and Europe shows that in any one season, a soil testing low in a nutrient often will not yield as well as a soil testing at an optimum level, no matter how much fertilizer is applied that year.

- Interpretation of soil test results and recommendations often becomes a matter of how to improve the fertility status of soils testing less than optimum. How much will be needed to change the soil from low to medium or high in that element? What will be the most economical level at which to maintain the nutrient status of the soil?

- With top-level management practices, yields increase and the probability of a response at any given soil test likewise increases.

- Wise use of soil testing incorporates a long-term approach to fertility management, in which site-specific soil test target levels are established for each field and nutrient management plans developed to reach and maintain the target levels.

The goal of soil testing is to help farmers achieve economical optimum yields while protecting the environment. The basic philosophy of soil test fertilizer recommendations is:

- Base them on soil test results;

- Recommend that lower-testing soils be built up to higher test levels by adding extra fertilizer;

- Apply maintenance amounts of plant nutrients to higher-testing soils to keep them there and to keep productivity high; and

- Do not apply specific nutrients to soils testing very high in these nutrients.

Individualized fertilizer recommendations use site- and grower-specific information, rather than laboratory-generated recommendations based on assumptions and generalizations. Computer programs are available that help personalize recommendations by considering the following:

- Soil Test Calibration Relevancy

How appropriate is the calibration used in the standard recommendation for the field in question? Unusual soil types, a different climate, no-till or ridge-till culture, crop variety, cropping history and field variability are examples of factors that could cause differences.

- Yield Potential

Yield potential determines the economic value of each percentage change in relative yield and may influence the shape of the calibration curve.

- Fertilizer Placement

Band placement often reduces lost yield as sub-optimal soil test levels are built to optimum levels because the short-term recovery of applied fertilizer by crop plants is improved. Some recommendation systems reduce the rate recommended when banding is used, compared to broadcast. However, rate studies have shown the optimum rate when banding is sometimes equal to or greater than the optimum broadcast rate. It is wise to build soil test levels to optimum regardless of placement method used.

- Farmer Financial Circumstances

The financial objective of farmers, like other investors with limited capital, is to maximize the return on the last dollar invested after considering all possible investment alternatives and their associated risks. Therefore, cash flow influences fertility management decisions.

- Uniform and Balanced Nutrient Distribution

Balance recommendations to ensure each nutrient is used efficiently.

- Land Tenure (Period of Time the Grower will Farm the Field)

Soil test phosphate and potassium are capital investments, and buildup costs should be amortized over the expected time of ownership or operation. The longer the period of time benefits will be accrued from buildup, the lower the cost of buildup becomes and the higher the optimum soil test level becomes. Landowners and operators, as well as the environment, benefit from the development of agreements in which the costs and returns of soil test buildup are equitably shared. Such agreements can help avoid the loss of productivity and accelerated erosion typical of run-down farms having impoverished soil fertility.

- Soil Test Buffer Potential

Soil test buffer potential is the quantity of fertilizer required to change the soil test level, and is usually expressed as pounds of P_2O_5 or K_2O required per ppm of soil test level change. Some low-pH and some high-pH soils fix applied phosphate readily, and increasing soil test phosphate is more costly, decreasing the optimum soil test level. Soil test phosphate and potassium levels are usually easier to change in sandier soils than on medium or fine-textured soils, except with very sandy soils, where potassium leaching becomes significant.

Recommendations When Levels Are High

Once soil tests are interpreted, possible approaches to a nutrient management plan may include the following:

- Sufficiency: Add necessary rates of deficient nutrients so yields are not limited in present crop.

- Build-Maintenance: Add enough of needed nutrient(s) to supply present crop need, and gradually increase soil supply to non-limiting level. Replace crop harvest–removed nutrients to keep plant nutrient levels at non-limiting levels.

If soil tests high in a plant nutrient, applying more of that nutrient is not recommended, at least for the current crop. This is especially true if there is an abundance of the nutrient present to the extent that there is almost no chance of response even if the nutrient was not applied for several years. However, some laboratories assign the value high to a level that points to little or no response to applications of that nutrient that year.

Failure to apply any of these nutrients will result in soil test depletion. Also, under some conditions, crops will respond profitably to a nutrient even with a high test. For example, on early-planted corn, the addition of N, P and K as a row application may produce response on soils testing high.

Fertilizer application when soils test in the high range is influenced substantially by the factors discussed in the section on individualization of recommendations. Maintenance in the high soil test category will be appropriate for some growers and sites but not for others

Sampling Procedures

Think about why you are sampling the soil. The goal is to estimate the capacity of the soil to provide adequate amounts of the necessary nutrients to meet the needs of the crop (or crops) to be grown. It should be clearly understood that soil testing does not measure the amount of nutrients in the soil. The test results can only be used in conjunction with a calibration curve that relates the laboratory analysis to a set of crop response data. Without the response (calibration) data, the laboratory results are meaningless. The samples should be collected in such a way as to best meet that goal. The sampling pattern should be set to best characterize the variability within the field.

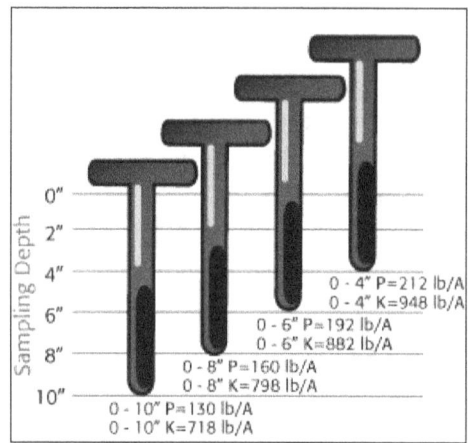

Effect of sampling depth on P and K soil test results.

Depth

Before sampling, check with the laboratory that will conduct the analysis to see what sampling depth is recommended. Sampling depth should be determined to represent the root zone the plant will draw from, but should also be consistent with the sampling depth used in developing the calibration data set to be used for interpreting the soil tests. Most soil test calibrations are based upon a 6- to 8-inch depth, most commonly 6 2/3 inches. In dry years, when it is difficult to push the probe into the ground, there is a danger of not getting the proper depth. Sampling too shallow will often lead to unusually high soil test results, because of the tendency for nutrients to become concentrated near the surface. Shallow sampling will thus overestimate the actual soil nutrient status and lead to underestimating fertilizer rates needed. This problem is even greater in reduced-tillage systems.

Uniformity of soil sampling depth is one of the most critical parts of soil testing, yet it is one of the most common sources of error. The figure at left illustrates an extreme example that emphasizes the problem. These sample results represent the difference in soil test P and K results for 4", 6", 8" and 10" sampling depths from Herman Warsaw's high-yield field, which produced a 370 bushel-per-acre corn yield in 1985. Though the numbers are not as dramatic, similar variation is common in any field, and is even more pronounced in reduced-tillage and no-till fields, where stratification of nutrients is common.

Pattern

Whether variable–rate nutrient application is planned or not, sampling the soil in an organized pattern is a good management practice. It helps ensure adequate representation of the entire field. Most agronomists recommend sampling on a pattern so that each sample represents about 2 ½ acres (one hectare) or less. At least one sample per acre is preferred, especially in areas receiving 25" or more of annual rainfall and in irrigated fields.

Sampling where Banded Fertilizer has been used

Banded fertilizer applications complicate the process of getting representative sampling. The recommendation is to take a number of samples between bands equal to eight times the distance (in feet) between bands. For example, if bands are 30 inches (2 ½ feet), there should be 20 samples (8 x 2 ½ = 20) collected between bands for each sample collected in the band.

Soil-Sampling Instructions

Important: Accurate soil analysis with meaningful interpretation requires properly taken samples. Follow all directions carefully and correctly. Sampling technique presents the greatest chance for errors in results. Laboratory analytic work will not improve the accuracy of a sample that does not represent the area.

1. Select the Proper Equipment

Collect samples using chrome plated or stainless steel sampling tubes or augers. Avoid galvanized, bronze or brass tools. Use clean, plastic buckets. Do not use galvanized or rubber buckets, as they will contaminate the samples.

2. When to Take Samples

Sampling can take place during any period of the year. However, it is best to sample a field at about the same time of year. Wait a minimum of 30 days to sample after applications of fertilizer, lime or sulfur.

3. Sample Area

Samples must be representative of the area you are treating. Most often, sampling by soil color is an acceptable method for dividing large fields into "like" areas. County ASCS aerial photographs can be used as a guide. Areas that differ in slope, drainage, past treatment, etc., should be sample separately. Sampling across dissimilar soil types is not recommended. And finally, the sample area should be large enough for special lime or fertilizer treatments.

Always remember to remove any surface debris prior to sampling.

Do Not Sample:

- Dead or back furrows
- Fence rows, old or new
- Old roadbeds, or near limestone gravel roads
- Terrace channels
- Wind breaks or snow fence lines
- Turn-rows
- Spill areas
- Fertilizer bands including anhydrous N
- Unusual or abnormal spots

4. Sample Depth

Sampling depth must remain consistent because many soils are stratified, and variation in depth will introduce errors into the analytical results. To test for soil stratification, sample through the soil profile, separately, 0" to 2", 2" to 4", 4" to 6", and 6" to 8". Remember to take the recommended number of cores per sample. The greater the difference in the analytical data between samples, the greater the degree of stratification.

5. Numbers of Cores and Acres per Sample

Various studies have shown that proper sampling requires at least 10 cores per sample, and sometimes 15 or more cores, depending on the nature of the soil and size of the area being sampled. A smaller number can introduce variability into the results from different sampling years. There is no rule for the number of acres to include in a single sample. This must depend on the local situation. However, the University of Illinois has long recommended that a single sample should represent no more than 5 acres.

6. Preparing Samples for Shipment

Thoroughly mix the randomly taken core samples in a plastic bucket, and remove a well-mixed composite sample (1/2 to 1 pint) from the mixture. Place it into the lab's sample bag, filling it to the "line." All samples taken for nitrogen analyses should be immediately air-dried, shipped early in the week, or shipped frozen.

Once the sample is in the bag, fold the top down to exclude air, roll it down to close, and fold the tabs. Write your sample ID designation (include grid sub-sample identification where applicable) and your customer's name on the bag where requested. Complete all the remaining information as required.

Sampling Pattern Options

The sampling pattern should be selected to best represent the field, accounting for known sources of variability (major soil type changes, past cropping patterns, etc.). A grid pattern is usually the best way to be sure the entire field is represented, but with the possibility of patterns developing from past nutrient applications, cropping effects and other uniform patterns, it is advisable to use a sampling scheme that avoids arranging sampling points in a straight line.

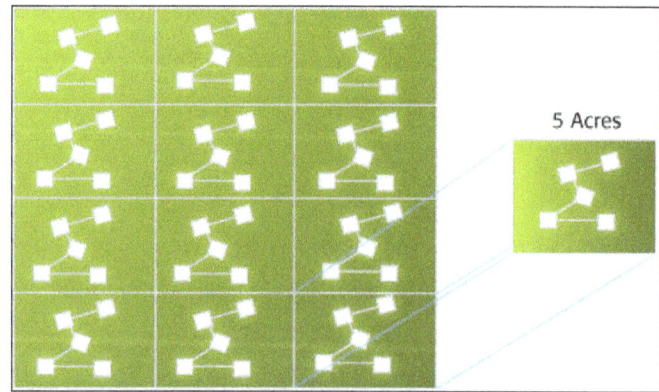

Area (Cell) Sampling Technique: Soil test values represent an area.

For conventional sampling, a common approach is to divide the field into cells of about 2½ to 5 acres, and collect five cores in a zigzag pattern within each cell to make up the sample. This area sampling method provides for fairly complete sampling of the field, and a good estimate of the need for a single uniform application rate to be applied to the entire field.

To better characterize the field for site-specific management and variable-rate application, point samples can be used to measure the variability across the field. Dividing the field into 2 ½-acre grids and collecting a sample for each cell, the grid lines help ensure a good spatial representation of the field that can be used to develop a nutrient map. Again, five cores should be collected, but they should be within a 10-foot radius of the center point for the sample. This provides nutrient information for the point, and the collection of data for all points in the field provides the basis of nutrient-variability maps. Several different interpolation schemes are used to estimate the nutrient levels across the field based upon the sample points. The more points, the more accurate the map, but there is a practical and economic limit to the sample density.

Grid Point Sampling Technique: Soil test values represent a point (Stratified Systematic Square Grid).

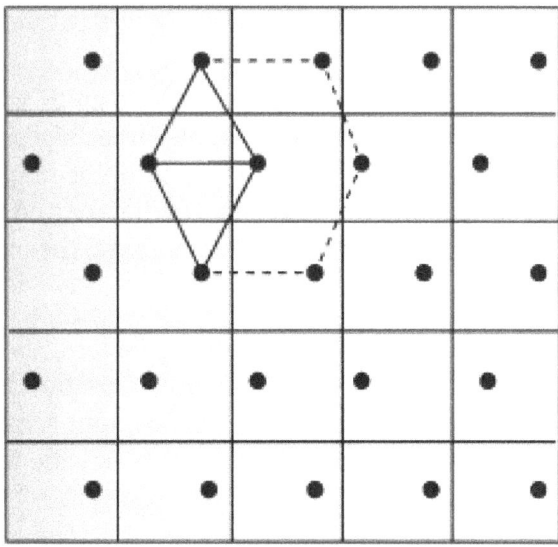

Stratified systematic sampling triangle, diamond or hexagon.

To avoid sampling bias caused by patterns in the field due to tillage, crop residue, fertilizer application, and other patterns associated with crop production, a staggered pattern can be used. It helps avoid the pattern bias, yet provides an organized sampling scheme to represent the entire field. This pattern can be set up by counting rows, using a measuring wheel, or using a global positioning satellite (GPS) navigation system. To gain the benefits of grid sampling, yet also the benefits of random sampling, the stratified systematic unaligned sampling pattern can be used to help avoid the effects of any patterns in the field.

Geo-referencing records. The GPS provides accurate positioning of the sample points, so that accurate geo-referenced maps of nutrient levels can be made with geographic information systems (GIS), and related to other data sets such as yield maps, soil survey, and remote sensing imagery. Even if GPS is unavailable, sample points should be referenced.

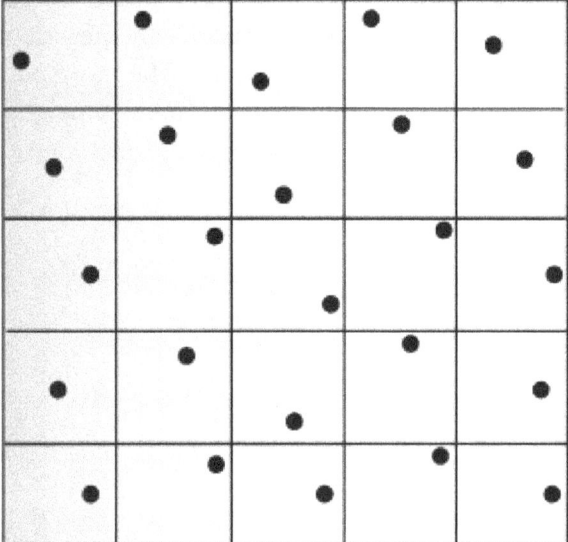

Stratified Systematic unaligned sampling.

Auxiliary Data Layers

Knowledge of specific sources of yield variability can be used to guide the sampling pattern. Additional samples may be taken to represent known wet spots, areas where cattle feedlots had previously been located, etc. Soil Survey maps, yield maps, topographic maps, aerial photographs and management histories are examples of auxiliary data layers that may be helpful in determining the best sampling pattern. If these data layers are in a GIS database, they may be used to help refine the recommendations for the field.

Soil survey maps are useful in determining major limiting factors, such as poor draining, steep slopes, and erosion. Soil survey data can be used to identify variation in soil organic matter, soil texture and other factors influencing changes in soil water content across the field and over time. This is important information to guide nutrient applications, pesticide rates, and other production inputs.

Sampling by Soil Type

Some agronomists prefer to set sampling patterns to reflect variation in soil types within the field. This plan requires a good soil survey map for the field, which may be obtained from the Natural

Resources Conservation Service (NRCS). Digital soil surveys being developed for many counties can be incorporated into the GIS database, making all of the data associated with soil types available as a part of the management tool package.

Where intensive, site-specific management is planned, it may be helpful to have a special Order 1 Soil Survey prepared for the field. The local NRCS office should be able to help identify a soil scientist who can prepare such a survey.

As with grid sampling, you will need to choose between area sampling (several cores taken at random points throughout the soil type boundary and mixed together for the sample) or point sampling (several cores collected within a few feet of specific sample points within the boundaries of each soil type).

If point sampling is used, the points can be geo-referenced so they can be related to other data sets or to future soil sampling.

The number of samples should be based on the known variability within the field. The number of cores per sample can also be chosen on that basis. Generally at least five, and preferably eight, cores per sample should be collected. The cores for each sample should be thoroughly mixed before being sent to the lab for analysis.

Soil Survey

Soil surveys are an important tool for nutrient management planning. They provide useful information for interpreting soil test results and predicting response to added nutrients. Most of the natural variability in soil nutrient levels and productivity is due to the characteristics documented in the soil survey. It is an excellent place to start in designing a sampling plan for nutrient management.

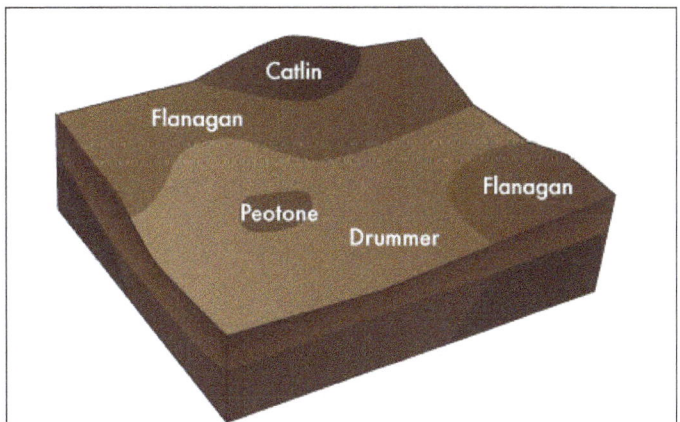

In this sampling plan, sample points are set to lie between the bounds of the different soil types, with care taken to avoid sampling on the transition between soil types.

These diagrams, from Bob McLeese, Illinois state soil scientist for Natural Resources Conservation Service (NRCS), illustrate a common problem with following a strict grid approach to sampling. The depression area on the topographic map appears on the soil survey map as Peotone-330. If a straight grid is used to establish sampling points, none of the points lies in the Peotone area, so it is missed entirely. In fact, of the 64 sample points in this field, up to 40 percent fall on boundaries between soil types.

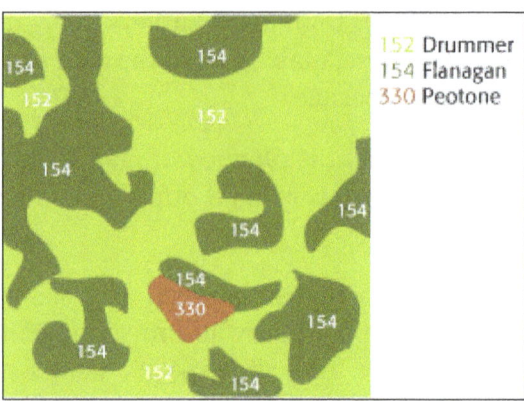

Soil survey

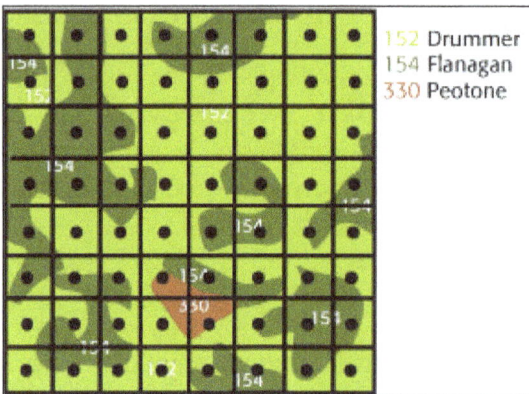

Straight grid over soil survey

By using a soil survey map or topographic map to "bias" the sampling and be sure the sample points are well within a given soil type, the influence of soil type and topography can be better taken into consideration when interpreting soil test values. While the relative importance of soil type of the soil test results in influence by many factors, it is helpful to avoid this "Peotone" problem whenever possible.

Whether sampling by soil type or by grid, the soil survey should be consulted in designing the sampling pattern to be used.

Smart Sampling or Biased Sampling

It is common sense, and good management, to adjust sampling patterns to help account for known sources of variability, such as topography, previous management patterns, old livestock lots or fencerows. These features can affect soil test levels and should be considered in determining sampling points. Even if a grid sampling pattern is used, it should be adjusted for known sources of nutrient variability. In some cases, you need to avoid these specific features. In other cases, it may be important to collect samples to adequately represent them.

Combinations

Many combinations of these different sampling patterns could be used. For example, grid sampling within soil types is a popular variation that gives some of the benefits of both systems. Select the pattern that will best represent the field. Remember the goal is to best represent the variability within the field. Design a pattern that will best do that. Even if variable-rate application is not planned, having the geo-referenced soil test record can be a valuable management resource. It also helps prepare for future implementation of variable-rate systems.

Choose a time that is convenient and allows adequate time to get results back from the lab and interpretations and recommendations made in time for the applications of nutrients. Sampling time is flexible, but it is important to sample at the same time each year if you intend to compare results from one year to the next. A few helpful guidelines:

- Be sure to record the date of sampling. Some recommendations may require adjustment factors for samples taken at different times of the year.

- Avoid midsummer, especially on sandy soils, where wetting and drying cause movement of salts and affect the pH.

- Sample before seeding or liming on acid soils where perennial forage crops will be planted.

- Avoid late-winter sampling on heavily textured soils. Freezing and thawing tend to release potassium and give unusually high soil test readings.

- Use October-to-December sampling for spring fertilizer applications and March to April for fall fertilizer applications. These periods tend to have the lowest testing variability.

Sampling under Different Tillage Systems

Different tillage systems provide different amounts and different depths of mixing of nutrients. Often, nutrients become stratified — or layered — in the soil profile. This can affect availability of nutrients to the plant, especially if moisture conditions limit root activity at any time during the growing season. For example, if nutrients accumulate in the top 3 to 4 inches of the root zone and the soil dries out in midsummer, the plant may become undernourished because of positional unavailability of the nutrients. That is, the supply is actually there, but inaccessible to the roots due to lack of moisture.

Moldboard Plow

Where a moldboard plow is used at least once every two or three years, nutrients and pH are uniformly distributed throughout the plow layer. For P, K and lime recommendations, samples should be taken to the plow depth — usually about 8 inches. Try to avoid collecting samples from the last year's fertilizer band.

Mulch Tillage

Some nutrient and pH stratification can be expected in mulch tillage systems, including chisel, disk and field cultivator systems. Sampling to a depth of about 8 inches, with care to avoid old rows and fertilizer bands, is recommended. Since mulch tillage also helps maintain moisture, this stratification is not necessarily a problem, and may result in concentrations of nutrients in small zones of varying pH, which may enhance nutrient uptake efficiency.

No-till

Where continuous no-till is practiced, distinct stratification of pH and nutrients is observed. Samples for routine P and K analysis should be taken to a depth of about 8 inches, again attempting to avoid crop rows and fertilizer bands. Stratification under no-till has not proven to be a problem in most cases. However, under drought stress, long-term no-till fields may become nutrient deficient in the lower part of the old plow layer. Monitoring the 4- to 8-inch depth, especially for K, may be helpful. Deep band placement of K is an effective means of overcoming this weather-related problem. Since lime is relatively immobile, recommendations for continuous no-till fields where lime is surface-applied should be based on a 4-inch sample depth. This also means that the amount of lime applied should be one-half that recommended for a conventionally tilled field at the same pH.

Identifying Missed Opportunities Through Intensive Sampling

More-intensive sampling can help identify missed fertilizer and crop profit opportunities in high-testing fields. Consider a central-Illinois field with an average soil test K level of 358 lb/acre. According to the University of Illinois Agronomy Handbook, this soil test is in the range where only maintenance fertilizer application would be needed. Based on a yield goal of 200 bu/acre corn and 60 bu/acre soybeans, the maintenance recommendation would be 134 lb/acre K_2O for the two-year rotation.

Sampling on a 1-acre grid reveals the spatial variability of the soil test level making up that average. Using the "buildup plus maintenance" fertilizer recommendation determined on the basis of the 1-acre cells instead of the field average, 47 acres show a need for buildup application of K, 30 acres need maintenance only, and 13 acres need no K applied. This means that the field-average approach (in this case, maintenance only) would put fertilizer on 13 acres that need none, and would miss the opportunity to supply needed "buildup" nutrients on 47 acres.

This field is representative of much of the eastern Midwest, where a long history of fertilizer use has resulted in field average soil test K levels in the adequate range, but where significant areas within the field still need buildup applications to reach or maintain optimum productivity. There is no way to determine the total fertilizer market potential represented by these areas unless detailed grid sampling is done. For most fields, that means sampling every 1 to 2 ½ acres, either on a uniform grid, or a modified grid that accounts for known sources of variability.

Organic Matter in Soil

Soil organic matter comprises all living soil organisms and all the remains of previous living organisms in their various degrees of decomposition. The living organisms can be animals, plants or micro-organisms, and can range in size from small animals to single cell bacteria only a few microns long.

Non-living organic matter can be considered to exist in four distinct pools:

- Organic matter dissolved in soil water.

- Particulate organic matter ranging from recently added plant and animal debris to partially decomposed material less than 50 microns in size, but all with an identifiable cell structure. Particulate organic matter can constitute from a few percent up to 25% of the total organic matter in a soil.

- Humus which comprises both organic molecules of identifiable structure like proteins and cellulose, and molecules with no identifiable structure (humic and fulvic acids and humin) but which have reactive regions which allow the molecule to bond with other mineral and organic soil components. These molecules are moderate to large in size (molecular weights of 20,000 - 100,000). Humus usually represents the largest pool of soil organic matter, comprising over 50% of the total.

- Inert organic matter or charcoal derived from the burning of plants. Can be up to 10% of the total soil organic matter.

When plant and animal debris is added to soil, it is broken down by macro- and micro-organisms, initially into particulate organic matter, and finally into humus. The raw materials can vary greatly in their resistance to breakdown. Woody organic substances like lignins are very resistant, while more simple compounds like sugars are readily utilised. Along the way, microbial populations increase. In the process they synthesise their own compounds which add to the diversity. In turn, these organisms die and are consumed by others.

Carbon dioxide is a by-product of this complex chain of processes (microbes breathe out CO_2 just like we do). Over half of the carbon added to soil is lost as CO_2 during breakdown. Because of their varying reactivity, the turnover times for these different carbon fractions varies from a few months to tens of thousands of years.

Measuring Soil Organic Matter

While living organisms, particularly the plants we grow, are of vital importance to us, it is the non-living organic matter that we measure as 'soil organic matter'.

The most common methods for measuring soil organic matter in current use actually measure the amount of carbon in the soil. This is done by oxidising the carbon and measuring either the amount of oxidant used (wet oxidation, usually using dichromate) or the CO_2 given off in the process (combustion method with specific detection).

Laboratories these days generally report results as soil organic carbon. Those that report as soil organic matter have usually measured carbon and converted to organic matter by multiplying by 1.72. However, this conversion factor is not the same for all soils, and it is more precise to report soil carbon rather than organic matter.

The Amount of Organic Matter in Soil

The amount of carbon (the measure of organic matter) in a soil depends on a range of factors, and reflects the balance between accumulation and breakdown. The main factors are:

- Climate - For similar soils under similar management, carbon is greater in areas of higher rainfall, and lower in areas of higher temperature. The rate of decomposition doubles for every 8 or 9 °C increase in mean annual temperature. Tasmanian agricultural areas have a mean annual temperature of 11-13 °C.

- Soil type - Clay helps protect organic matter from breakdown, either by binding organic matter strongly or by forming a physical barrier which limits microbial access. Clay soils in the same area under similar management will tend to retain more carbon than sandy soils. Hence the sandy sodosols of the northern midlands have less carbon than the clay loam ferrosols of the north west regardless of management.

- Vegetative growth - The more vegetative production the greater are the inputs of carbon. Also, the more woody this vegetation is (greater C:N ratio), the slower it will breakdown. So, the crop system can strongly affect carbon concentrations.

- Topography - Soils at the bottom of slopes generally have higher carbon because these areas are generally wetter and have higher clay contents. Poorly drained areas have much slower rates of carbon breakdown.

- Tillage - Tillage will increase carbon breakdown. However, the impact of tillage is generally outweighed by the effect of management on the amount of carbon grown and returned to the soil. An exception to this is where tillage leads to increased erosion.

Table: Organic carbon concentrations % in Tasmanian soils subject to different management

Soil (depth)	Pasture	Intermittent cropping	Frequent cropping
Ferrosol (0-150 mm) Red soil on basalt	6.4	4.9	3.8
Dermosol (0-75 mm) Cressy clay loam	7.0	4.3	4.2
Sodosol (0-150 mm) Sandy soil over clay	2.7	2.3	1.8

Importance of Soil Organic Matter

A fertile and healthy soil is the basis for healthy plants, animals, and humans. And soil organic matter is the very foundation for healthy and productive soils. Understanding the role of organic matter in maintaining a healthy soil is essential for developing ecologically sound agricultural practices. But how can organic matter, which only makes up a small percentage of most soils, The reason is that organic matter positively influences, or modifies the effect of, essentially all soil properties. That is the reason it's so important to our understanding of soil health and how to manage soils better. Organic matter is essentially the heart of the story, but certainly not the only part. In addition to functioning in a large number of key roles that promote soil processes and crop growth, soil organic matter is a critical part of a number of global and regional cycles.

It's true that you can grow plants on soils with little organic matter. In fact, you don't have to have any soil at all. (Although gravel and sand hydroponic systems without soil can grow excellent crops, large-scale systems of this type are usually neither economically nor ecologically sound.) It's also true that there are other important issues aside from organic matter when considering the quality of a soil. However, as soil organic matter decreases, it becomes increasingly difficult to grow plants, because problems with fertility, water availability, compaction, erosion, parasites, diseases, and insects become more common. Ever higher levels of inputs—fertilizers, irrigation water, pesticides, and machinery—are required to maintain yields in the face of organic matter depletion. But if attention is paid to proper organic matter management, the soil can support a good crop without the need for expensive fixes.

The organic matter content of agricultural topsoil is usually in the range of 1–6%. A study of soils in Michigan demonstrated potential crop-yield increases of about 12% for every 1% organic matter. In a Maryland experiment, researchers saw an increase of approximately 80 bushels of corn per acre when organic matter increased from 0.8% to 2%. The enormous influence of organic matter on so many of the soil's properties— biological, chemical, and physical—makes it of critical importance to healthy soils. Part of the explanation for this influence is the small particle size of the well-decomposed portion of organic matter—the humus. Its large surface area–to–volume ratio means that humus is in contact with a considerable portion of the soil. The intimate contact of humus with the rest of the soil allows many reactions, such as the release of available nutrients into the soil water, to occur rapidly. However, the many roles of living organisms make soil life an essential part of the organic matter story.

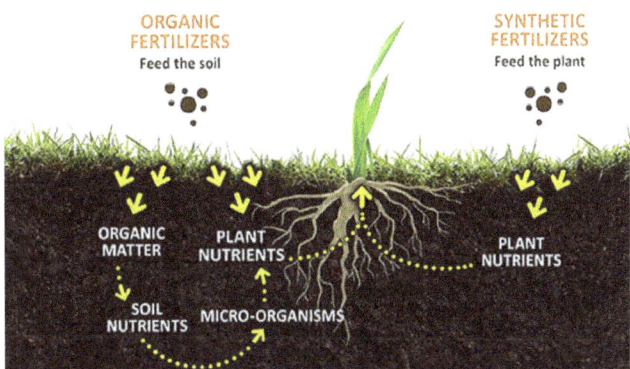

Adding organic matter results in many changes modified from Oshins and Drinkwater

Plant Nutrition

Plants need eighteen chemical elements for their growth—carbon (C), hydrogen (H), oxygen (O), nitrogen (N), phosphorus (P), potassium (K), sulfur (S), calcium (Ca), magnesium (Mg), iron (Fe), manganese (Mn), boron (B), zinc (Zn), molybdenum (Mo), nickel (Ni), copper (Cu), cobalt (Co), and chlorine (Cl). Plants obtain carbon as carbon dioxide (CO_2) and oxygen partially as oxygen gas (O_2) from the air. The remaining essential elements are obtained mainly from the soil. The availability of these nutrients is influenced either directly or indirectly by the presence of organic matter. The elements needed in large amounts—carbon, hydrogen, oxygen, nitrogen, phosphorus, potassium, calcium, magnesium, sulfur—are called macronutrients. The other elements, called micronutrients, are essential elements needed in small amounts. (Sodium [Na] helps many plants grow better, but it is not considered essential to plant growth and reproduction.)

Nutrients from Decomposing Organic Matter

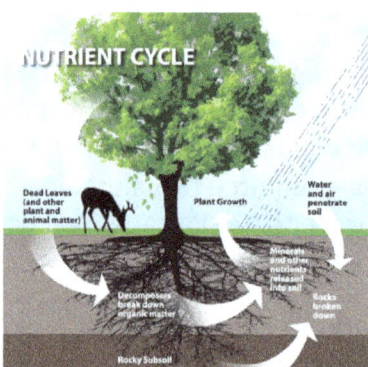

The cycle of plant nutrients

Most of the nutrients in soil organic matter can't be used by plants as long as those nutrients exist as part of large organic molecules. As soil organisms decompose organic or mineral forms that plants can easily use. This process, called mineralization, provides much of the nitrogen that plants need by converting it from organic forms. For example, proteins are converted to ammonium (NH_4^+) and then to nitrate (NO_3^-). Most plants will take up the majority of their nitrogen from soils in the form of nitrate. The mineralization of organic matter is also an important mechanism for supplying plants with such nutrients as phosphorus and sulfur and most of the micronutrients. This release of nutrients from organic matter by mineralization is part of a larger agricultural nutrient cycle.

Addition of Nitrogen

Bacteria living in nodules on legume roots convert nitrogen from atmospheric gas (N_2) to forms that the plant can use directly. A number of free-living bacteria also fix nitrogen.

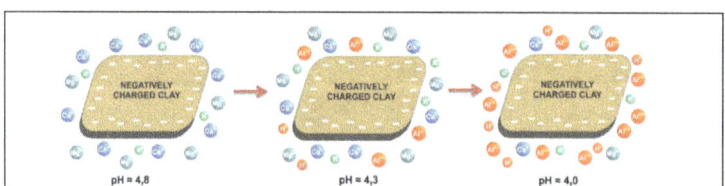

Cations held on negatively charged organic matter and clay

Storage of Nutrients on Soil Organic Matter

Decomposing organic matter can feed plants directly, but it also can indirectly benefit the nutrition of the plant. A number of essential nutrients occur in soils as positively charged molecules called cations (pronounced cat-eye-ons). The ability of organic matter to hold on to cations in a way that keeps them available to plants is known as cation exchange capacity (CEC). Humus has many negative charges. Because opposite charges attract, humus is able to hold on to positively charged nutrients, such as calcium (Ca^{++}), potassium (K^+), and magnesium (Mg^{++}). This keeps them from leaching deep into the subsoil when water moves through the topsoil. Nutrients held in this way can be gradually released into the soil solution and made available to plants throughout the growing season. However, keep in mind that not all plant nutrients occur as cations. For example, the nitrate form of nitrogen is negatively charged (NO_3^-) and is actually repelled by the negatively charged CEC. Therefore, nitrate leaches easily as water moves down through the soil and beyond the root zone.

Clay particles also have negative charges on their surfaces, but organic matter may be the major source of negative charges for coarse and medium-textured soils. Some types of clays, such as those found in the southeastern United States and in the tropics, tend to have low amounts of negative charge. When those clays are present, organic matter may be the major source of negative charges that bind nutrients, even for fine-textured (high-clay-content) soils.

Protection of Nutrients by Chelation

Organic molecules in the soil may also hold on to and protect certain nutrients. These particles, called "chelates" (pronounced key-lates) are by-products of the active decomposition of organic materials and are smaller than the particles that make up humus. In general, elements are held more strongly by chelates than by binding of positive and negative charges. Chelates work well because they bind the nutrient at more than one location on the organic molecule. In some soils, trace elements, such as iron, zinc, and manganese, would be converted to unavailable forms if they were not bound by chelates. It is not uncommon to find low-organic-matter soils or exposed subsoils deficient in these micronutrients.

Other Ways of Maintaining Available Nutrients

There is some evidence that organic matter in the soil can inhibit the conversion of available phosphorus to forms that are unavailable to plants. One explanation is that organic matter coats the surfaces of minerals that can bond tightly to phosphorus. Once these surfaces are covered, available forms of phosphorus are less likely to react with them. In addition, humic substances may chelate aluminum and iron, both of which can react with phosphorus in the soil solution. When they are held as chelates, these metals are unable to form an insoluble mineral with phosphorus.

Beneficial Effects of Soil Organisms

Soil organisms are essential for keeping plants well supplied with nutrients because they break down organic matter. These organisms make nutrients available by freeing them from organic molecules. Some bacteria fix nitrogen gas from the atmosphere, making it available to plants. Other organisms dissolve minerals and make phosphorus more available. If soil organisms aren't present and active, more fertilizers will be needed to supply plant nutrients.

A varied community of organisms is your best protection against major pest outbreaks and soil fertility problems. A soil rich in organic matter and continually supplied with different types of fresh residues is home to a much more diverse group of organisms than soil depleted of organic matter.

This greater diversity of organisms helps insure that fewer potentially harmful organisms will be able to develop sufficient populations to reduce crop yields.

Soil Tilth

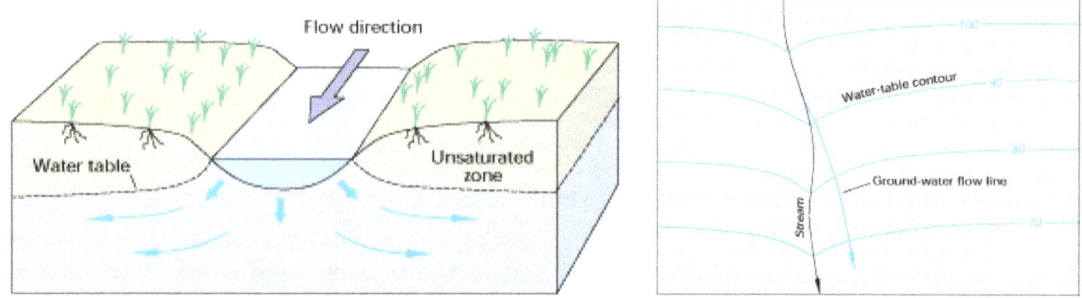

Changes in soil surface and water-flow pattern when needs and crusts develop

When soil has a favorable physical condition for growing plants, it is said to have good tilth. Such a soil is porous and allows water to enter easily, instead of running off the surface. More water is stored in the soil for plants to use between rains, and less erosion occurs. Good tilth also means that the soil is well aerated. Roots can easily obtain oxygen and get rid of carbon dioxide. A porous soil does not restrict root development and exploration. When a soil has poor tilth, the soil's structure deteriorates and soil aggregates break down, causing increased compaction and decreased aeration and water storage. A soil layer can become so compacted that roots can't grow. A soil with excellent physical properties will have numerous channels and pores of many different sizes.

Studies on both undisturbed and agricultural soils show that as organic matter increases, soils tend to be less compact and have more space for air passage and water storage. Sticky substances are produced during the decomposition of plant residues. Along with plant roots and fungal hyphae, they bind mineral particles together into clumps, or aggregates. In addition, the sticky secretions of mycorrhizal fungi—beneficial fungi that enter roots and help plants get more water and nutrients—are important binding material in soils. The arrangement and collection of minerals as aggregates and the degree of soil compaction have huge effects on plant growth. The development of aggregates is desirable in all types of soils because it promotes better drainage, aeration, and water storage. The one exception is for wetland crops, such as rice, when you want a dense, puddled soil to keep it flooded.

Organic matter, as residue on the soil surface or as a binding agent for aggregates near the surface, plays an important role in decreasing soil erosion. Surface residues intercept raindrops and decrease their potential to detach soil particles. These surface residues also slow water as it flows across the field, giving it a better chance to infiltrate into the soil. Aggregates and large channels greatly enhance the ability of soil to conduct water from the surface into the subsoil.

Most farmers can tell that one soil is better than another by looking at them, seeing how they work up when tilled, or even by sensing how they feel when walked on or touched. What they are seeing or sensing is really good tilth. It shows that soil differences can be created by different management strategies. Farmers and gardeners would certainly rather grow their crops on the more porous soil depicted in the photo on the right.

Since erosion tends to remove the most fertile part of the soil, it can cause a significant reduction in crop yields. In some soils, the loss of just a few inches of topsoil may result in a yield reduction of 50%. The surface of some soils low in organic matter may seal over, or crust, as rainfall breaks down aggregates and pores near the surface fill with solids. When this happens, water that can't infiltrate into the soil runs off the field, carrying valuable topsoil.

Large soil pores, or channels, are very important because of their ability to allow a lot of water to flow rapidly into the soil. Larger pores are formed in a number of ways. Old root channels may remain open for some time after the root decomposes. Larger soil organisms, such as insects and earthworms, create channels as they move through the soil. The mucus that earthworms secrete to keep their skin from drying out also helps to keep their channels open for a long time.

Protection of the Soil against Rapid Changes in Acidity

Acids and bases are released as minerals dissolve and organisms go about their normal functions of decomposing organic materials or fixing nitrogen. Acids or bases are excreted by the roots of plants, and acids form in the soil from the use of nitrogen fertilizers. It is best for plants if the soil acidity status, referred to as pH, does not swing too wildly during the season. The pH scale is a way of expressing the amount of free hydrogen (H+) in the soil water. More acidic conditions, with greater amounts of hydrogen, are indicated by lower numbers. A soil at pH 4 is very acid. Its solution is ten times more acid than a soil at pH 5. A soil at pH 7 is neutral—there is just as much base in the water as there is acid. Most crops do best when the soil is slightly acid and the pH is around 6 to 7. Essential nutrients are more available to plants in this pH range than when soils are either more acidic or more basic. Soil organic matter is able to slow down, or buffer, changes in pH by taking free hydrogen out of solution as acids are produced or by giving off hydrogen as bases are produced.

Stimulation of Root Development

Corn grown in nutrient solution with (right) and without (left) humic acids. Photo by
R. Bartlett. In this experiment by Rich Bartlett adding humic acids to a nutrient solution increased
the growth of tomatoes and corn as well as the amount and branching of roots

Microorganisms in soils produce numerous substances that stimulate plant growth. Humus itself has a directly beneficial effect on plants. The reason for this stimulation has been found mainly to be due to making micronutrients more available to plants—causing roots to grow longer and have more branches, resulting in larger and healthier plants. In addition, many soil microorganisms produce a variety of root-stimulating substances that behave as plant hormones.

Darkening of the Soil

Organic matter tends to darken soils. You can easily see this in coarse-textured sandy soils containing light-colored minerals. Under well-drained conditions, a darker soil surface allows a soil to warm up a little faster in the spring. This provides a slight advantage for seed germination and the early stages of seedling development, which is often beneficial in cold regions.

Protection Against Harmful Chemicals

Some naturally occurring chemicals in soils can harm plants. For example, aluminum is an important part of many soil minerals and, as such, poses no threat to plants. As soils become more acidic, especially at pH levels below 5.5, aluminum becomes soluble. Some soluble forms of aluminum, if present in the soil solution, are toxic to plant roots. However, in the presence of significant quantities of soil organic matter, the aluminum is bound tightly and will not do as much damage.

Organic matter is the single most important soil property that reduces pesticide leaching. It holds tightly on to a number of pesticides. This prevents or reduces leaching of these chemicals into groundwater and allows time for detoxification by microbes. Microorganisms can change the chemical structure of some pesticides, industrial oils, many petroleum products (gas and oils), and other potentially toxic chemicals, rendering them harmless.

References

- Soil_physics: researchgate.net, Retrieved August 9, 2019

- Soil -texture: the-compost-gardener.com, Retrieved June 7, 2019

- Soil -texture-definition-classes-and-its-determination, soil-texture: soilmanagementindia.com

- Zobraz_cast.pl?cast, opory, eknihovna: mendelu.cz, Retrieved February 15, 2019

- Efu-soil-testing: cropnutrition.com, Retrieved May 21, 2019

- Efu-soil-sampling: cropnutrition.com, Retrieved June 19, 2019

- Soil-organic-matter, soil-management, land-management-and-soils,agriculture: tas.gov.au, Retrieved January 7, 2019

- Why-Soil-Organic-Matter-Is-So-Important, Organic-Matter-What-It-Is-and-Why-It-s-So-Important, Text-Version, Building-Soils-for-Better-Crops-3rd-Edition, Books, Learning-Center: sare.org, Retrieved February 3, 2019

Chapter 4

Soil Chemistry

The study of the chemical characteristics of soil is known as soil chemistry. It deals with the study of soil reaction such as oxidation reaction, soil ph, soil salinity, soil sodicity, etc. The factors that affect soil chemistry are mineral composition, environmental factors and organic matter. This chapter discusses in detail the diverse aspects related to soil chemistry.

The chemistry of the soil is also very important property as this will determine what will grow and how well it will grow. One of the most important chemical properties of a soil is its acidity or alkalinity, often stated as the pH of the soil. The pH of the soil ranges from about 3 to 8. Below 5.5 the soil is quite acid. Above pH 7 the soil is alkaline. Soil with a pH in the range of 5.5 to 7 tends to be the most flexible and a wide range of plants can thrive within this pH range. Once the pH drops below 5.5, firmly into the acidic range, there is only a limited range of plants that like this level of acidity and can tolerate these acidic conditions. Once the pH is above 7.0, the soil tends to be colonised by a limited range of lime-loving plants.

Soils in wet climates and certainly those developed on acidic rocks, such as granite, will tend to be acid. Soils in high rainfall areas tend to be acid because the rainfall leaches the soil of many of its nutrients which otherwise help to keep the pH higher. Acid soils can be improved by adding lime to soil. This is a common agricultural practice where farm soils need to be maintained at a pH from 5.5 to 7 in order to grow a wide range of crops.

Soils formed in dry climates are often alkaline, i.e., with a pH above 7. Here there is a lack of rainfall to flush the nutrients out of the soil and they stay within the soil. In some dry conditions also evaporation may lead to deep-lying nutrients being brought to the surface. This can lead in some dry-climate soils to excessive amounts of salts being brought into the root zone. In extreme cases this can lead to salinization, in which the soils contain too many salts, which can prevent the growth of many crops.

Soil Colloids

Colloid: A particle, either mineral or organic, with a diameter of 0.1 to 0.001 μm. Because of their small size, colloids go into suspension in a solution—they float around for great lengths of time without settling out. Clay particles and soil organic matter are common examples of soil colloids.

Importance

Colloids have properties that are important in soil chemistry. For example, because of their small size they have a high relative surface area that has a charge, so they can adsorb cations. This is key for Cation Exchange Capacity, but also for maintaining the structure of the soil (binding particles together) and for its water-holding capacity (higher concentration of colloids means greater ability to hold water).

Soil Solution

Water in the soil is referred to as the soil solution because it contains dissolved materials (cations and ions) as well as suspended colloids of clay and organic matter While plants tend to get their nutrients from the soil solution, the solution does not contain sufficient nutrients at any one time to last the life of the plant. Usually these nutrients are replenished from the pool of exchangeable nutrients. Still more nutrients are held in what is called the stable pool (bound up in solid form as minerals or organic matter).

Cation Exchange Capacity (CEC) and Base Saturation

CEC

a) Definition CEC: is a measure of the ability of the soil to adsorb cations. Plants are primarily able to take up the ionic form of nutrients via their roots. Many of these nutrients are taken up as cations (remember, these are positive ions). Most soils have at least some ability to hold onto cations at negatively charged sites, called exchange sites, on soil particles. The cations are held loosely to the edges (adsorbed) such that they can be easily replaced with similarly charged cations. The total amount of the cations that the soil can hold adsorb is the cation exchange capacity (CEC).

b) Measurement: CEC is measured as milliequivalents (meq) per 100g of soil or centimoles (cmol) per kg. These are actually two ways of expressing the same numbers.

c) Factors influencing CEC:

   1. Amount and type of clay Higher amounts of clay in the soil mean higher CEC. Certain kinds of clay (smectites, montmorillonite) have higher CEC than others (such as kaolinite).

   2. Amount of organic matter: Higher amounts of organic matter in the soil mean higher CEC

   3. pH-dependent CEC: Clay minerals and organic matter have a CEC that varies with pH. As pH increases, so do the number of negative charges on the clay or organic matter particles, and thus so does the CEC.

Base Saturation

a) Definitions Base saturation refers to the percentage of CEC sites that are occupied with bases (usually Ca^{2+}, Mg^{2+}, K^+ and Na^+) instead of ions that make the soil acidic (H+ or Al^{3+}). Base saturation is often expressed as a percent. The term exchangeable bases usually refers to the Ca^{2+}, Mg^{2+}, K^+ and Na_+ adsorbed to CEC sites.

b) Significance Soils with high base saturations are considered more fertile because many of the "bases" that contribute to it are plant nutrients. Usually the base saturation is 100 percent when the pH is above about 6.5. Since rainfall tends to leach bases out of the soil, areas with higher rainfall tend to have lower base saturations than areas with lower rainfall, unless the parent material is high in bases (such as limestone).

Anion Exchange Capacity (AEC)

   1. Definition: While positively-charged cations adsorb to negatively-charged sites, the opposite is true for negatively-charged anions: They adsorb to sites with a positive charge. This is anion exchange capacity, AEC. Nutrients that are usually supplied by anions are nitrogen (as NO_3), phosphorus (as HPO_4^{2-}), sulfur (as SO_4^-), chlorine (as Cl^-), boron (as $B_4O7_2^-$) and molybdenum (MoO_4^-).

   2. Measurement: AEC is measured as milliequivalents (meq) per 100g of soil or centimoles (cmol) per kg.

   3. pH-dependent AEC: Most clay particles only have negative exchange sites, so they have CEC in neutral and high pH conditions and sometimes AEC at low pH. Soil organic matter has both negative and positive exchange sites; it usually has CEC and may have AEC in very low pH (2 or lower) conditions.

   4. Nutrient leaching: Because there is generally little adsorption of anions, many (particularly nitrates) are easily leached down through the soil with rain or excess irrigation. This can lead to groundwater contamination, which can even happen in organic farming if the N is not well managed.

pH

pH stands for "potential of hydrogen" and it is expressed as the negative of the log of the concentration of hydrogen (H+) ions. It is given as a number between 0 and 14. (Pure water is neutral with a pH of around 7.) In acidic soils (pH < 7), H^+ ions predominate. In alkaline soils (pH > 7), OH- ions predominate. Soils with pH of 7 are neutral.

Table: Soil Reaction and pH

REACTION	pH	REACTION	pH
Ultra acid	< 3.5	Neutral	6.6-7.3
Extremely acid	3.5 – 4.4	Slightly alkaline	7.4-7.8
Very strongly acid	4.5-5.0	Moderately alkaline	7.9-8.4
Strongly acid	5.1-5.5	Strongly alkaline	8.5-9.0
Moderately acid	5.6-6.0	Very strongly alkaline	>9.0
Slightly acid	6.1-6.5		

- Effect of pH on Nutrient Availability and Uptake

Although pH does not directly affect plants, it does affect the availability of different nutrients to plants. As we've seen in the CEC and AEC sections above, nutrients need to be dissolved in the soil solution before they can be accessed by plants. The soil pH changes whether a nutrient is dissolved in the soil solution or forms other less-soluble compounds (e.g., calcium compounds in high pH soils with high calcium carbonate concentrations), or if dissolved is then susceptible to leaching (e.g., nitrate).

- pH and Soil Microbes

Soil microbes have reduced activity in low pH soils. This can cause them to take much longer to release necessary nutrients, such as N, P, and S, from organic matter.

Acidity

Acidity refers to the condition of the soil when the exchange sites on soil colloids (collectively called the exchange complex) are dominated by hydrogen (H^+) and aluminum (Al^+) ions. As described above, these soils have pH <7.

- Distribution of Acid

Soils Acidic soils usually occur where rainfall leaches the cations out of the soil over time. In the U.S. there is a fairly strong correlation between precipitation and pH, with soils receiving more than about 30 inches of annual precipitation having a pH < 6.

- Problems Associated with Acidity

Aluminum toxicity: Aluminum becomes more available when pH is pH <6 and especially <4.75, and can be toxic to plants.

- Acid soils and liming

Lime (calcium carbonate) is added to acid soils to raise the pH. Calcium (Ca^{2+}) replaces hydrogen and aluminum on the exchange sites.

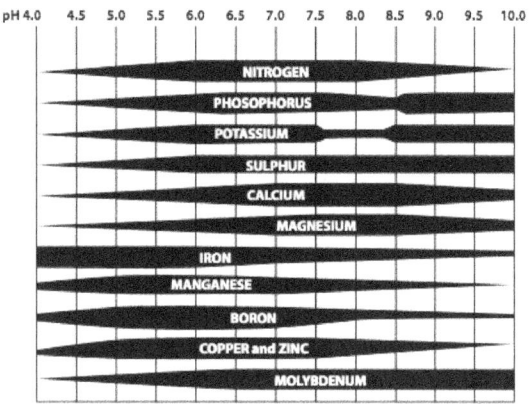

Nutrient availability at different ph values. Maximum availability is indicated by widest part of bar

Alkaline, Saline and Sodic Soils

1. Alkalinity and acidity: Soils that vary from a neutral pH have varying degrees of alkalinity (pH > 7) or acidity (pH < 7). The mean soil pH in the U.S. is around 6.4.

2. Salinity: Soils that have excess soluble salts in the soil solution have varying degrees of salinity

3. Sodicity: Soils that specifically have excess sodium in the soil solution are called sodic

4. Alkalinity Soils in arid and semi-arid areas can lack enough rainfall to leach cations, especially calcium (Ca^{2+}), magnesium (Mg^{2+}), potassium (K^+) and sodium (Na^+), from the soil. These cations bind many of the CEC sites, blocking hydrogen (H^+) ions from binding and making the soil alkaline. This can also happen if irrigating with water high in calcium bicarbonate or magnesium bicarbonate.

Salinity

A soil containing sufficient soluble salts (these salts include Mg^{2+}, Na^+, Ca^{2+}, chloride (Cl^-), sulfate (SO_4^{2-}), bicarbonate (HCO_3^-) and carbonate (CO_3^{2-}). Saline soils mainly occur in dry areas, again, where there is not enough precipitation to leach the salts from the soil, so the salts build up over time. In order for there to be salts in the soil, there must be a source for them. Some come from former ocean floors that were under ancient seas but are now exposed. Some parent material (rocks from which the soil was formed) also may release salts, such as carbonate from limestone or sodium from feldspar.

Some salts are toxic to plants and others bind so tightly to water that the plants cannot access it. In addition, it can be difficult for non-saline water to infiltrate saline soils, so it may be necessary to add gypsum to the water to aid infiltration.

Sodicity

A soil containing sufficient exchangeable sodium to adversely affect crop production and soil structure under most conditions of soil and plant types. Many saline soils are also sodic, although not necessarily. Sodium is toxic to plants. It also causes soil particles to disperse (separate), which causes cracking and sealing of the soil surface, leading to poor soil structure and decreased water intake.

Sodic soils can be reclaimed with a two-step process. First the sodium is flushed from CEC sites by adding amendments high in calcium (such as lime, gypsum, or dolomite) or by adding sulfur followed by lime. (The sulfur is converted to sulfuric acid by microbial activity. The sulfuric acid then reacts with lime to free calcium.) In either case, the Ca^{2+} ions replace the $Na+$ cations, freeing the Na^+ in the soil solution. The second step is to leach out the sodium ions by irrigating in excess of what the plant needs.

Quantitative Definitions

Specifically, alkaline, saline, and sodic soils are defined as such:

a) Alkaline soil: Has a pH of > 8.5 or with an exchangeable sodium percentage (ESP, that is, the percent of the CEC occupied just by sodium) greater than 15%. Soils at this ESP contain sufficient sodium to interfere with the growth of most crop plants.

b) Saline soil: Soil salinity is determined by measuring the electrical conductivity (EC) of a saturated paste of soil: if the EC is greater than 4 dS/m (decisiemens per meter), the soil is classified as saline. However this is a rough range: salt-sensitive plants can be affected at half this EC and highly tolerant plants can handle up to about twice this EC.

c) Sodic soil: A soil in which the ESP is at least 15%. The amount of sodium in the soil may also be expressed by the Sodium Adsorption Ratio (SAR), which reflects the degree to which the CEC sites in the soil are occupied by sodium instead of other cations. A soil with a SAR greater than 13 is considered sodic. An ESP of 15% is roughly equivalent to a SAR of 13.

d) Saline-sodic soil: A soil containing both high soluble salts in general and high sufficient exchangeable sodium in particular. The ESP is at least 15%, the EC of the soil solution is >4 dS/m, and the pH is usually < 8.5.

Soil as a Medium for Plant Growth

Nutrient uptake Processes

Imagine you are a tiny creature trying to move around in the soil. You are surrounded by millions of pores of all sizes and shapes, shaped and blocked by particles of organic matter and minerals. The surfaces of these particles are chemically active, adsorbing ions and organic molecules all around you. You start to learn your way around, but your microenvironment changes with each wet-and-dry cycle and freeze-and-thaw cycle. Sometimes it's not a physical process but a biological one that rearranges the structure of your little world, like a burrowing animal that tunnels through. In short, you live in a constantly changing soil ecosystem that has numerous barriers to the movement of organisms and chemicals.

In terms of soil fertility we are particularly interested in the physical component of the soil ecosystem. For a nutrient to be available for the plant to take up it must meet two criteria: 1) it must be in the proper chemical form to pass the root membrane; and 2) it must be available at the root surface.

Nutrients move through the soil to plant roots in three ways:

- Root interception.

- Mass flow.

- Diffusion.

Each nutrient will have one or more of these methods of movement depending on its chemical form (including how strongly they are adsorbed by mineral and organic matter particles) and soil physical and chemical conditions (including the concentration of the nutrient in the soil).

Root Interception

Plant roots are constantly expanding (opening up blocked pores as they do so), growing from areas of depleted nutrients (e.g., because of prior plant uptake) to regions where nutrients are more concentrated.

Although many plants, such as cereals and other grasses, have a very extensive root system, they contact less than 5% of the soil volume. The root interception mechanism is very valuable, however, because root growth can extend the root into areas where mass flow and diffusion then take over. For example, a root could grow within a few millimeters of some soil phosphorus hot spot. Although the root does not technically bump into the nutrient and intercept it, the root is close enough for diffusion to occur and the phosphorus to move into it. In some cases, the presence of mychorrhizal fungi increases the nutrient-absorption capacity of root systems. Root interception allows for uptake of some calcium, magnesium, zinc, and manganese.

Mass Flow

Growing plants are continually taking up water from the soil profile, a process driven by transpiration (loss of water from the plant via stomata on the leaves). Dissolved in the soil water are soluble nutrients. These nutrients are transported along with the water to the root surface. Nutrients, such as nitrogen as nitrate and sulfur as sulfate, that are held very weakly by soil particles readily move along with the water. But nutrients, such as phosphorus as orthophosphate, that are strongly adsorbed to the soil particles are not able to reach roots by mass flow. Mass flow allows for the uptake of most of a plant's nitrogen, calcium, magnesium, sulfur, copper, boron, manganese and molybdenum.

Diffusion

Diffusion is the movement of ions along a gradient from a high concentration to a lower concentration, until the ions are evenly distributed. For example, imagine you have a tank of water with a removable barrier in the middle. On one side of the barrier you pour ink, while the other side stays pure water. If you remove the barrier very slowly you will see the ink and water mix as the molecules move from an area of high concentration (the inky side) to an area of low concentration (the pure water side). Similarly, nutrients move from areas of high concentration in the soil solution to areas of lower concentration. This is a very slow process, but it is the dominant mechanism of movement for phosphorus and potassium, which are strongly adsorbed on the soils and present in low concentrations in the soil solution.

Soil Reaction

The degree of acidity or alkalinity of a soil is called soilreaction. It is an indicator of the acidity or alkalinity and is measured in pHscale. It is the most outstanding characteristics for plant growth factors because it determines the availability of plant nutrients and the amount of toxic substance in the soil.

Microorganisms and higher plants respond markedly to soilreaction because it tends to control so much of their chemical environment.It is used in diagnosing the fertility as well as productivity of soils.

Types of Soil Reaction

One of the outstanding physiological characteristics of the soil solution is its reaction. Soil reaction influences many physical and chemical properties of soil. The growth and activity of plant and soil organisms depend on soil reaction and the factors associated with it.

There can be three types of soil reaction as follows:

1. Acidity

    Soil acidity is common in regions where precipitation is high enough to leach appreciable amounts of exchangeable bases from the surface layers of the soil. The two adsorbed cations such as Hydrogen and Aluminium are largely responsible for soil acidity. The acid soil is generally found in humid region. The factors which will help in the release and removal of bases will help in the development of acidity of soil. If the hydrogen (H^+) ion becomes more than hydroxyl (OH^-) ion in the soil solution, the soil becomes acidic.

    A highly acidic soil may have pH 4.5 and low calcium and magnesium, high solubility of iron, manganese, aluminium etc., but low availability of nitrogen and phosphorus. The activity of microorganism responsible for nitrification is adversely affected in acid soil. Generally limes are used for reclamation of acid soil.

2. Alkalinity

    The soil that contains absorbed sodium to interfere with the growth of most crop plant is known as alkali soil. The amount of exchangeable sodium in great quantities in the soil makes the soil alkalinity. The sodium ion easily displaced the calcium ion from clay colloid and makes the sodium mixed clay particles. This sodium is converted into sodium hydroxide by hydrolysis as per the following reaction:

$$2\,Na^+ + CO_3^{-2} + 2\,H_2O \rightarrow 2\,Na^+ + 2\,OH^- + H_2CO_3$$

The OH⁻ ion thus formed increases the soil pH.

3. Neutrality

   In those areas, where the soil contain hydrogen and hydroxyl ion almost in equal quantities, the soils are neutral in character.

Soil pH

pH is the negative logarithm of hydrogen ion concentration of soil solution and it is usually written as –

$$pH = \log \frac{1}{\left(H^+\right)} \text{ or } pH = -\log 10\ CH^+$$

pH scale is used to measure the acidity or alkalinity of a soil solution (or other solution). This scale runs from 0 pH to 14 pH. In this scale, the 7 units level is known as neutral i.e. neither acidic nor alkaline. Pure water has a pH-7.0. All values below pH 7.0 denotes acidity and the values above pH 7.0 represents alkalinity. The degree of acidity increases as pH decreases below pH 7.0. Soil showing pH 5 is ten times more acidic than showing pH 6.0. Likewise, the degree of alkalinity increase as we go higher from pH 7.0. The alkalinity at pH 9.0. unit is ten times more than that pH 8.0 units.

Factors Controlling Soil Reaction

There are some factors that control soil reaction are as follows:

1. Nature of Soil Colloid

   Soil colloid influences soil reaction to a very great extent. Soil colloids when dominated by adsorbed hydrogen (H^+) ion, the reaction of soil becomes acidic. On the other hand, soil colloid when dominated by hydroxyl (OH^-) ion, the reaction of soil becomes alkalinity.

2. Nature of Ion

   The soil that contains more hydrogen ion than hydroxyl ions becomes acidic in reaction. When the aluminium ions are present in the soil, they react with water to liberate hydrogen ions, which increases the soil acidity

   $$AI^{+++} + 3\,HOH \rightarrow AI(OH)_3 + 3\,H^+$$

3. Percentage Base Saturation

   A low percentage base saturation of soil means soil acidity. In humid areas, the basic elements have been leached down from the soil, the percentage base saturation decreases much below 80 and they become acidic in reaction. If the percentage of base saturation is above 80 and at 90, then they become neutral in reaction and alkaline reaction respectively.

4. Rainfall

    Rainfall plays important role in determining the soil reaction. The soils that are developed in high rainfall areas, becomes acidic in nature due to leaching of some nutrients such as calcium (Ca^{++}), magnesium (Mg^{++}) etc. from soil solution. So leaching encourages the development of soil acidity. On other hand, the soils that are developed in low rainfall areas, becomes alkaline in nature.

5. Fertilizers

    The continual use of fertilizers is responsible for a marked change in soil pH. Acid forming fertilizers such as Ammonium sulphate, Urea, Ammonium nitrate etc. when applied in the soil in large quantities makes the soil acidic. On the other hand, basic fertilizers such Sodium nitrate, Basic slag etc. makes the soil alkaline.

Influence of Soil Reaction on the Availability of Nutrients

Soil reaction affects the availability of nutrients to plants. Soil reaction affects indirectly in availability of nutrients to plants as soil organism do not function well in acid and alkali soil. Soil organism do their function at their best within a pH range 6.0-7.5.

The influence of soil reaction on the availability of nutrients to plant is as follows:

1. Nitrogen:

    Nitrogen is most important nutrient for plants. Plant absorbs nitrogen in the form of Ammonium (NH_4^+) and Nitrate (NO_3^-). Out of these two forms, plant absorbs most of their nitrogen in the form of nitrate (NO_3^-). The availability of nitrate nitrogen depends on the activity of nitrifying bacteria's. The microorganism responsible for nitrification are most active when the soil pH is between 6.5 and 7.5.

    The activity of nitrifying bacteria is adversely affected if pH falls below 5.5 and more than 9.0. The activity of nitrogen fixing bacteria (e.g. Azotobacter) falls down at below soil pH 6.0. In acidic condition, the decomposition of organic matter, the main source of nitrogen, is also slow down.

2. Phosphorus:

    Phosphorus is an essential constituent of every living cells and for nutrition of plant and animal. The availability of phosphorus depends on the soil pH. In strongly acidic soil (pH 5.0 or less), iron, aluminium, magnesium and other bases remains in soluble form and phosphorus reacts with these bases are converted into insoluble form and become unavailable to plant.

    $$Al^{3+}\ H_2PO_4^- + 2H_2O \rightleftharpoons 2H^+ + Al(OH)_2\ H_2PO_4$$
    $$\underset{\text{(Soluble)}}{\phantom{Al^{3+}\ H_2PO_4^-}} \qquad\qquad \underset{\text{(Insoluble)}}{}$$

In acid soils, phosphorus becomes available to plants by anion exchange with the hydroxyl anion (OH^-)

$$AI(OH)_2 H_2PO_4 + OH^- \rightleftarrows AI(OH)_2 + H_2PO_4^-$$
$$\underset{(\text{Insoluble})}{} \qquad\qquad \underset{(\text{Soluble})}{}$$

One anion (OH) has been exchanged for another ion Phosphorus ($H_2 PO_4^-$) become available after liming in the acid soil.

Phosphorus fixation takes place even when the soil is alkaline. Phosphate ion combine with calcium ion and calcium or magnesium carbonate and form insoluble calcium or magnesium carbonate.

$$\underset{(\text{soluble})}{Ca(H_2PO_4)_2} + \underset{(\text{Adsorbed})}{2Ca^{++}} \rightleftarrows \underset{(\text{Insoluble})}{Ca_3(PO_4)_2} + 4H^+$$

$$Ca(H_2PO_4) + 2CaCO_3 \rightleftarrows \underset{(\text{Insoluble})}{Ca_3(PO_4)_2} + 2CO_2 + 2H_2O$$

The availability of phosphorus remains highest at a soil pH between 6.5-7.5.

3. **Potassium:**

   Potassium is an essential element for the development of chlorophyll. The availability of potassium does not influence by soil reaction to any great extent. In acid soil, potassium is lost through leaching. Application of lime for reclamation of acid soil result in an increase in potassium fixation of soils and the potassium remains in the soil in the form of non-availability.

4. **Sulphur:**

   Sulphur is an important element for oil seeds, cruciferae, sugar and pulse crop. The availability of sulphur is not affected by soil reaction. In acid soil, it is more soluble and is subjected to loss by leaching.

5. **Calcium:**

   Calcium as calcium pectate is an important constituent of cell wall and requires in large amounts for cell division. Acid soils are poor in calcium. In alkali soil (pH not exceeding 8.5), the availability of calcium remains high. The availability of calcium decreases when soil pH is above 8.5.

6. **Magnesium:**

   Magnesium is an essential constituent of chlorophyll. Acid soils are poor in magnesium. In alkali soil (pH not exceeding 8.5), the availability of magnesium remains high. The availability of magnesium decreases when soil pH is above 8.5.

7. **Manganese:**

   Manganese is an essential constituent of chlorophyll and also formation of oils and fats. Soil pH has decided influence in the availability of manganese. At high pH values, all cations are unfavourably affected. Over liming or a naturally high pH is associated with deficiencies of manganese and such conditions occur in nature in many of the calcareous soils of West Bengal.

8. Iron:

Iron is necessary for the synthesis of chlorophyll. In very acid soil, there is relative abundance of ions of iron. Iron deficiency of plant due to high pH is not uncommon. At high pH i.e. in alkali soils, ferrous (Fe_2^+) ion is converted to ferric (Fe_3^+) and precipitated as Ferric oxide (Fe_2O_3). The availability of iron increases as the pH of soil decreases.

9. Zinc:

Soil pH affects the availability of Zinc. The zinc deficiency occurs on soils that are slightly acidic to neutral. High pH reduces the availability of zinc by precipitating zinc as zinc hydroxides.

10. Boron:

Boron occurs in most soil in extremely small quantities. The availability and utilization of boron is determined to a considerable extent by soil pH. Boron is most soluble under acid condition. It apparently occurs in acid soil in part as boric acid and this is readily available to plants. The high soil pH causes boron deficiency in plants forming complex compound. A specific Ca : B is required for every crop. When calcium level is high, boron content should be high and if no, plant will show boron deficiency.

11. Copper:

Copper is an essential constituent of enzyme. In very acid soil, there is relative abundance of copper. The solubility of copper decreases as pH increases particularly in sandy soils. The decrease in solubility with increasing the soil pH may be result of precipitation of copper in the form of cupric oxide (CuO). The oxidized state of copper i.e. hydroxides or hydrous oxides is insoluble. Copper deficiency is induced by heavy liming and excessive application of nitrogen and phosphorus.

12. Molybdenum:

Molybdenum is a constituent part of the enzyme, nitrate reductase. Molybdenum availability is significantly dependent on soil pH. It is quite unavailable in strongly acid soil and becomes available by liming of acid soil.

Oxidation Reaction

In soil systems, rapid changes in moisture content strongly affects soil aeration status. In addition to changes in bulk conditions, there are microsites or zone where diffusion of oxygen is restricted, such as small pores filled with water and the interior of aggregates where oxygen consumption is more rapid than oxygen diffusion. In areas were there is better access to soil air and pockets of easily decomposed organics, and intense decomposition can deplete available oxygen and produce a number of redox-active organic compounds.

From a global perspective redox is an important aspect of energy and carbon transfer. Carbon reduction and its subsequent oxidation fuel the biological world. Carbon reduction or gain in electrons is often called photosynthesis by the uninitiated and the reverse oxidation reaction is called respiration when applied to humans and other large animals. Microbial respiration accompanies or is synonymous with mineralization.

Regardless of your perspective, redox reactions are important aspects of soil chemistry. Redox reactions change the speciation and solubility of many elements, create new compounds and alter the biochemistry of soils. In a complex mixture such as soils the interpretation of redox relationships is difficult. Since the dynamics of soil oxygen which drives the changes in redox potential are rapid, equilibrium may not be attained. From a thermochemistry view point redox is not at equilibrium, because all of the energy yielding compounds by definition contain excess free energy and are unstable with respect to carbon dioxide and water. Processes which reduce oxygen levels and decrease redox potentials are driven by microbial consumption of oxygen. Conditions necessary for lowering redox potentials include, a source of decomposable organic materials (energy source), a population of microbes capable of utilizing this energy source for metabolism, and a restriction on the resupply of oxygen. These requirements are not uniformly distributed in soils and sediments. Thus, redox reactions and redox potentials are not uniform throughout the soil matrix. In fact, redox potentials are highly variable and therefore are best used as an indication or relative status of the soil.

Thermochemistry background

The Gibbs free energy (G) may be defined in differential form as:

$$dG = -Sd\,T - V\,dP - w'$$

Where w' is defined as the useful work in a chemical system (non-pressure, volume work) in our case this will be the electrical work of the system.

At constant T and P we find:

$$dG = -w'$$

Also remembering that at equilibrium dG = 0 and therefore the useful work or electrical work in the system is also zero at equilibrium. In an electro-chemical system the work is derived from the transfer of electrons from one compound to another and is equal to the potential and the charge transferred w'= EdQ, where E is the potential and Q is the quantity of charge transferred. In a chemical reaction where n electrons are transferred per mole of reactant, the electrical work is nFE where E is the emf of the system, F is the Faraday constant. Therefore:

$$d\,G = -nFE$$

and

$$d\,G^o = -nFE^o$$

From thermodynamics it can also be shown that:

$$dG = dG^o + R\,T\,\ln Q$$

Q is the reaction quotient.

Equating $dG = -nFE$, $dG^\circ = -nFE^\circ$ and $dG = dG^\circ + RT \ln Q$ we find $E = E^\circ - \dfrac{RT}{nF} \ln Q$ which is the familiar Nernst Equation,

$$E = E^\circ - \frac{RT}{nF} \ln Q$$

Other useful relations are derived from $dG^\circ = -nFE^\circ$ and the relationship $dG^\circ = -RT \ln K^\circ$ to obtain:

$$-nFE^\circ = -RT \ln K^\circ$$

Rearranging and substituting gives:

$$\log K^\circ = 16.9 n E^\circ$$

When E is expressed in volts. Solving for E° evaluating constants and specifying temperature as 298 °K expressing the result in millivolts gives:

$$E^\circ = \frac{59.2}{n} \log K^\circ$$

The Nernst Equation is the basis for the measurement of redox potential in soils and sediments. Replacing E with Eh and E° with E_h° and writing the reaction as a reduction produces the familiar equation used for soil redox potential (E_h)

$$E_h = E_h^\circ - \frac{0.059}{n} \log \frac{(\text{reduced species})}{\text{oxidized species}(H^+)^m}$$

In aqueous systems the bounds of Eh is dictated by the stability of water. In oxidizing systems the oxidation of water to yield oxygen and protons via the following reaction is the upper boundary for Eh. Since the reaction yields protons, the stability is a function of pH.

$$2H_2O \rightarrow O_2 + 4H^+ + 4e^-$$

$$E^\circ = 1.229 \text{ volts}$$

Using the Nernst Equation evaluated at 25°C gives the following:

$$E_h = E^\circ + \frac{0.059}{4} \log \left(\frac{P_{O2}(H^+)^4}{(H_2O)^2} \right)$$

for conditions where the activity of water is taken to be unity the equation reduces to:

$$E_h = E^\circ + \frac{0.059}{4} \log P_{O_2} + 0.059 \log(H^+)$$

$$E_h = E^\circ + 0.0148 \log P_{O_2} - 0.059 \text{ pH}$$

$$E_h = 1.229 + 0.0148 P_{O_2} - 0.059 \text{ pH}$$

Water oxidation depends on both pH and oxygen pressure. If an arbitrary oxygen pressure is chosen, than the plot of Eh vs pH will have a negative slope of 0.059 V or 59 mV per pH unit and an intercept of E° (1.229 volts). Note that if (H^+) is 1 mole/liter (i.e. pH = 0), and P_{O_2} is 1 atmosphere standard conditions are met and Eh = E°, which is the intersection of the line with the vertical axis in Eh-pH plots.

At the other extreme of reducing conditions in aqueous systems, hydrogen ion reduction to hydrogen gas is the lower stability of the system. As written below for as an oxidation potential, the reaction is:

$$H_{2g} \rightarrow 2H^+_{aq} + 2e^-$$

$$E^\circ = 0.000 \, volts$$

$$E_h = E^\circ + \frac{0.059}{2} \log\left(\frac{(H^+)^2}{P_{H_2}}\right)$$

$$E_h = E^\circ - 0.0295 \log P_{H_2} - 0.059 \, pH$$

but since $E^\circ = 0$ for the Hydrogen half cell,

$$E_h = -0.0295 \log P_{H_2} - 0.059 \, pH$$

At standard conditions where P_{H2} = 1 atmosphere and pH = 0, E_h = E° = 0. For a given hydrogen pressure the Eh decreases with a slope of 0.059 volts per pH unit and has an intercept of 0.000

Thus the stability limits for E_h are bounded by hydrogen and oxygen gas evolution from water. Limits of 1 atmosphere of oxygen and/or hydrogen are not realistic, however they give the bounds for the system. Other values for Eh given oxygen levels of 0.2 atmospheres can easily be calculated and will be parallel to the 1 atmosphere lines. At pH = 0, 0.0148 log PO2 = -1.229 is the oxygen pressure required for Eh to equal 0.

In soil and sediment systems reasonable bounds for pH are between about 4 and 9, therefore the Eh - pH relationships of soils are defined in the regions bounded by these values.

Note that in for standard conditions of 1 atmosphere of oxygen and pH = 0, E° = 1.229 volts. Since pe = E_h / 0.059 or 20.83. The slope of the E_h vs pH line is -0.059 therefore, as pH increases one unit E_h decreases 0.059 volts and pe + pH = a constant or 20.83. Similarly for the hydrogen water couple, pe + pH = 0.

Pourbaix Diagrams

These diagrams are an extension of the other equilibrium diagrams we have discussed this quarter. In this case the plots are values of E or pE in relation to pH. pH is chosen as the independent variable because oxidation reduction reaction are strongly affected by pH and this is more convenient than partial pressure diagrams involving P_{O_2}, of P_{H_2}. Hydrogen ions or hydroxide ions are often involved in the reactions and are conveniently covered by pH. As an example lets look at the diagram for iron superimposed on the water stability graph discussed in McBride.

$$Fe^{3+} + e^- \rightleftarrows Fe^{2+} \quad E° = 0.771V; \quad pE° = 13.02$$

$$Fe(OH)_3(s) \rightleftarrows Fe^{3+} + 3(OH^-) \quad pK_{sp} = 39.294$$

$$Fe(OH)_2(s) \rightleftarrows Fe^{2+} + 2(OH^-) \quad pK_{sp} = 15.096$$

The value for $Fe(OH)_3$ is that for soil ferric hydroxide.

Since equation $Fe^{3+} + e \rightleftarrows Fe^2 \quad E° = 0.771V; \quad pE \quad 13.0$ is independent of pH, it represents a line parallel to the pH axis.

$$E = E° - 0.059 \log \frac{(Fe^{2+})}{(Fe^{3+})}$$

Equation $E = E° - 0.059 \log \frac{(Fe^{2+})}{(Fe^{3+})}$ contains the activity ratio of Fe2+ to Fe3+, Fe concentration is not needed, however the straight line with an intercept at 0.771 volts implies that the activity ratio of the two iron species is always = 1.

Ferric iron level and pH will determine the line separating Fe^{3+} and $Fe(OH)_3$. Redox reactions are not involved and the line is parallel to the Eh axis. Choosing a Fe^{3+} value will determine the position of line A separating Fe^{3+} and $Fe(OH)_3$.

$$Fe^{3+} = \left(\frac{K_{sp}}{K_W^3}\right)(H^+)^3$$

Taking logs of both sides and substituting - 3 pH for 3 log (H^+), $10^{-39.294}$ for K_{sp} and 10^{-14} for K_w gives:

$$\log(Fe^{3+}) = 2.706 - 3pH; \quad \text{or} \quad pH = 0.902 - \frac{1}{3}\log(Fe^{3+})$$

If Fe^{3+} is taken to be 1 mM, separation between Fe^{3+} and $Fe(OH)_3$ is drawn at pH 1.902.

At the intersection of the line derived from equation $\log(Fe^{3+}) = 2.706 - 3pH; \quad \text{or} \quad pH = 0.902 - \frac{1}{3}\log(Fe^{3+})$ and the redox relation for Fe^{2+} and Fe^{3+}, solid phase $Fe(OH)_3$ can be reduced to Fe^{2+}. The redox equilibrium is derived by summing equations and adding 3 H^+ ions to each side of equation $Fe(OH)_3(s) \rightleftarrows Fe^{3+} + 3(OH^-) \quad pK_{sp} = 39.294$ to give:

$$3H_2O + Fe^{2+} \rightleftarrows Fe(OH)_{3(s)} + 3H^+ + e^-$$

This couple has an E° of + 0.9306 volts and has the form:

$$E_h = 0.9306 - 0.177\,pH - 0.059\log(Fe^{2+})$$

If the concentration of Fe²⁺ is taken as 1 mM, then the equation is:

$$E_h = 1.108 - 0.177\,pH$$

At the pH where our assumed iron concentration will cause precipitation of ferrous hydroxide, the relationship changes to: $Fe(OH)_3 + H^+ + e^- \Leftrightarrow Fe(OH)_2 + H_2O$ with an E^o of -0.66 volts

$$E_h = 0.166 - 0.059\,pH$$

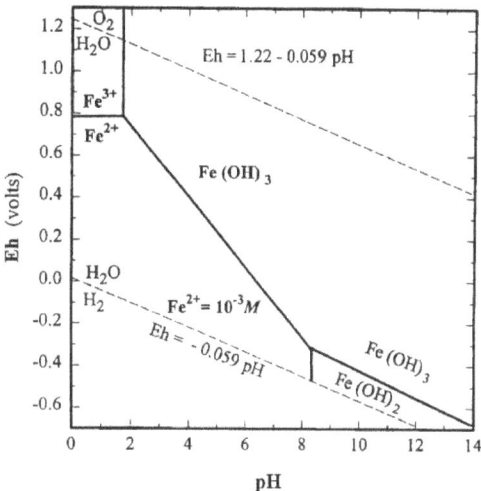

Pourbaix diagram for iron in relation to E_h and pH

Soil pH

Soil pH is determined by the concentration of hydrogen ions (H$^+$). It is a measure of the soil solution's (soil water together with its dissolved substances) acidity and alkalinity, on a scale from 0 to 14. Acidic solutions have a pH less than 7, while basic or alkaline solutions have a pH greater than 7. By definition, pH is measured on a negative logarithmic scale of the hydrogen ion concentration [H$^+$], i.e., pH = -log [H$^+$]. Therefore, as hydrogen ion concentration (and acidity) goes up, pH value goes down. Also, because pH is a logarithmic function, each unit on the pH scale is 10 times more acidic than the unit above it. For example, a pH 6 solution has a 10 times greater concentration of H$^+$ ions than a pH 7 solution, and a 100 times higher concentration than a pH 8 solution.

Soil pH is influenced by both acid and baseforming cations (positively charged dissolved ions) in the soil. Common acid-forming cations are hydrogen (H$^+$), aluminum (Al^{3+}), and iron (Fe^{2+} or Fe^{3+}), whereas common base-forming cations include calcium (Ca^{2+}), magnesium (Mg^{2+}), potassium (K$^+$) and sodium (Na$^+$).

Most agricultural soils in Montana and Wyoming have near-neutral to basic conditions with average pH values of 6.5 to 8. This is primarily due to the presence of base-forming cations associated with carbonates and bicarbonates found naturally in soils and irrigation waters. Due to relatively low precipitation amounts, there is little leaching of base-forming cations, resulting in pH values greater than 7.

There some areas in Montana and Wyoming with acidic soils. Acidic conditions occur in soil with parent material high in elements such as silica (rhyolite and granite), high levels of sand with low buffering capacities (ability to resist pH change), and in regions with higher amounts of precipitation. High precipitation causes leaching of base-forming cations and lowering of soil pH. Naturally acidic soils are most commonly found west of the continental divide or in high elevation areas, in areas where soils were formed from acid forming parent material, forest soils, mining sites containing pyritic (iron and elemental sulfur [S°]) minerals, and a few other isolated locations. Soil acidity in the seeding zone is becoming a problem on some cropland soils because of N fertilization.

Nutrient Availability

A soil's ability to hold and supply nutrients is related to its cation and anion exchange capacities, the number of parking spaces for nutrients on soil particles. Cation and anion exchange capacities are influenced by soil pH. As described in Plant Nutrition and Soil Fertility, cation and anion exchange capacity are largely determined by the charge of the soil particles and SOM. Soils with high amounts of clay and/or organic matter typically have higher cation exchange capacity (CEC), that is, are able to bind more cations such as calcium or potassium than more silty or sandy soils. They also have greater buffering capacity.

Soil pH affects nutrient availability because the H^+ ions take up space on the negative charges along the soil surface, displacing nutrients. The effect on nutrient availability depends on the size and charge of the nutrient molecules and whether or not they can be lost to leaching.

The metal nutrients (e.g., copper [Cu], iron, manganese [Mn], zinc [Zn]) are small molecules when dissolved in water with 2 to 3 positive charges, thus a high charge to size ratio. They bind strongly to the surface of soil particles. At high pH (i.e., basic, low H^+ concentration), these metal ions stick so tightly they are not readily found in soil solution and thus are less available for plant uptake. At low pH (i.e., acidic, high H^+ concentration), fewer can stick to the soil surface, making them more available for plant uptake.

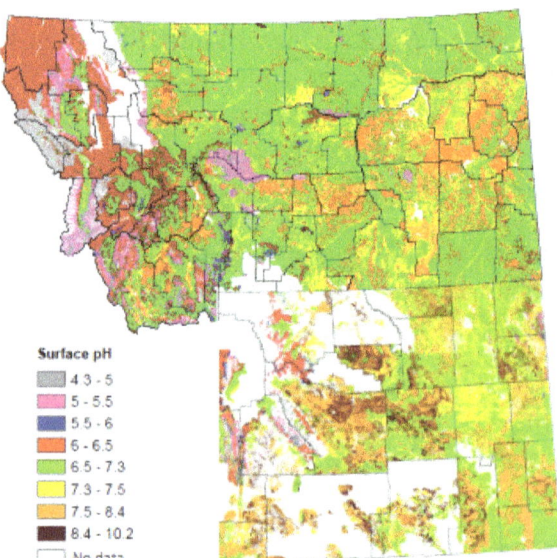

Soil surface horizon pH (generally 4 to 6-inch depth) in Montana and Wyoming

Sulfur (S) and the base-forming cations (Ca^{2+}, Mg^{2+}, K^+, and Na^+) are relatively large molecules. Like a large electrostatically charged balloon does not stick well to a wall, these large molecules do not stick tightly to soil particles. Therefore, even at high pH (low H^+ concentration), they easily come off of the soil particle and enter soil solution. At low pH they are displaced by H^+, and may not be plant available because they have been lost from the soil through leaching or uptake. Nitrate (NO_3^-) is equally available across soil pH levels because it doesn't bond much to soil.

In general, nitrogen (N), potassium, calcium, magnesium and sulfur are more available within soil pH 6.5 to 8, while boron (B), copper, iron, manganese, nickel (Ni), and zinc are more available within soil pH 5 to 7. At pH less than 5.5, high concentrations of H^+, aluminum and manganese in soil solution can reach toxic levels and limit crop production. Phosphorus is most available within soil pH 5.5 to 7.5.

Agronomic Concerns

In addition to the effects of pH on nutrient availability, and aluminum and manganese toxicity, individual plants and soil organisms also vary in their tolerance to basic and acid soil conditions. Neutral conditions appear to be best for crop growth. However, optimum pH conditions for individual crops vary. Some crop varieties are being developed to tolerate lower pH and higher aluminum levels.

Table: Optimal pH ranges for common crops in Montana and Wyoming.

Crop	Soil pH
Alfalfa	6.2 - 7.5
Barley	5.5 - 7.0
Dry bean	6.0 - 7.5
Corn	5.5 - 7.0
Oat	5.5 - 7.0
Pea	6.0 - 7.0
Potato	5.0 - 5.5
Sugar beet	6.5 - 8.0

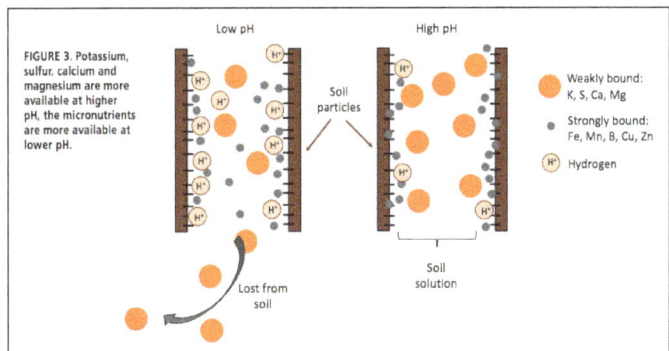

Potassium, sulfur, calcium and magnesium are more
available at higher pH, the micronutrients are more available at lower pH

Soil microorganism activity is greatest near neutral conditions, but optimal pH ranges vary for each type of microorganism. Microbial activity is considerably reduced at pH 5 and below. Moreover, certain 'specialized' microorganisms, such as nitrifying bacteria (convert ammonium [NH_4^+]

to nitrate [NO_3^-]) and nitrogen-fixing bacteria associated with many legumes, generally perform poorly when soil pH falls below 6. For example, alfalfa (a legume) grows best in soils with pH levels greater than 6.2, conditions in which its associated nitrogen-fixing bacteria grow well too. Potatoes grow well at soil pH 6.5, but the potato scab organism also thrives at that pH. Scab is greatly reduced at pH less than 5.2, which potatoes tolerate, but production requires higher fertilization to compensate for reduced nutrient availability at such low pH. In contrast, fungi generally thrive at low pH, so fungal diseases are more common in acidic soils. Finally, pesticide effectiveness and residual (carry-over) is an issue in acidic soils (9, Washington). When soil pH is extremely acidic or basic, pH modifications may be needed to obtain optimal growing conditions for specific crops.

Managing Soil pH

To manage soil pH, the addition of amendments, fertilization and tillage practices, SOM levels and crop selection should all be considered. The longevity of soil pH change brought about by management greatly depends on the treatment. Changes can occur within a season or last for decades.

Amendments

Sulfur

A common amendment used to lower the pH of basic soils is sulfur. Elemental sulfur is oxidized by microbes to produce sulfate (SO_4^{2-}) and H^+, causing a lower pH. Ferrous sulfate ($FeSO_4$) and aluminum sulfate ($Al_2[SO_4]_3$) can also be used to lower pH, not due to sulfate, but because of the addition of acidic cations (Fe_{2+}, Al_{3+}). Application rates for these amendments vary depending upon product properties (particle size, oxidation rate) and soil conditions (original pH, buffering capacity, minerals present). Because calcium carbonate ($CaCO_3$) consistently buffers soil to pH values near 8, soils high in calcium carbonate would need larger quantities of sulfur amendments to lower pH than generally economical. An unpublished study by Agvise Laboratories, Inc., found 230 lb S/1,000 ft2 (5 ton/acre) reduced soil pH from 8.0 to 7.5, and 115 lb gypsum/1,000 ft2 (2.5 ton/acre) had no impact on soil pH. At \$1 per pound sulfur, amending with sulfur might be worthwhile for a market garden, but certainly not for large scale crop production.

Lime

A common method for increasing soil pH is to lime soils with calcium carbonate, calcium oxide (CaO), calcium hydroxide (Ca[OH]$_2$), or calcium containing by-products such as sugar-beet lime. The liming material reacts with carbon dioxide and water in the soil to yield bicarbonate (HCO_3^-) and hydroxide (OH$^-$), which take H^+ and aluminum (acid-forming cations) out of solution, thereby raising the soil pH. The benefits are varied and depend on the soil pH level reached.

Companies supplying lime amendments are required to state the lime score (which is also called effective neutralizing value [ENV]), calcium carbonate equivalent (CCE), and particle size on their label. Lime score is a quality index used to express the effectiveness of liming materials for neutralizing soil acidity and is based on purity, particle size, and percent dry matter. Chemical purity is represented by CCE which compares the liming material to pure calcium carbonate. As CCE increases, the acid neutralizing power in the lime increases. Particle size is measured as the mesh size (number of screen wires per inch) through which ground lime will fall; increasing mesh size corresponds with smaller mesh openings. Fine sized lime (mesh size of 40 or greater) will react more effectively and quickly in the soil, whereas coarser sized lime will dissolve more slowly and remain in the soil for a longer period of time.

Table: Management practices that change soil pH.

Decrease	Increase
Elemental sulfur	Lime
Ammonium based nitrogen fertilizer (e.g., urea, 11-52-0)	Nitrate based nitrogen fertilizer (e.g., ammonium nitrate [34-0-0])
Leaf and stem harvest	Large quantity plant material left on field
Legumes on soil surface	Legumes in mature root zone
SOM to buffer	
Tillage to mix layers	

Table: The benefits of increasing pH of acidic soils.

Soil pH change	Effects
6.1 -› 6.5	Improve soil structure, reduce crusting, and reduce power need for tillage
5.6 -› 6.0	Increase soil microbial activity; increase rhizobia health for nitrogen-fixation and other mycorrhizal assisted crops (legumes and barley); increase plant nutrient availability; as above
5.1 -› 5.5	Reduce aluminum, H^+, and manganese toxicity; as above
< 5.1	Few crops can produce if not limed

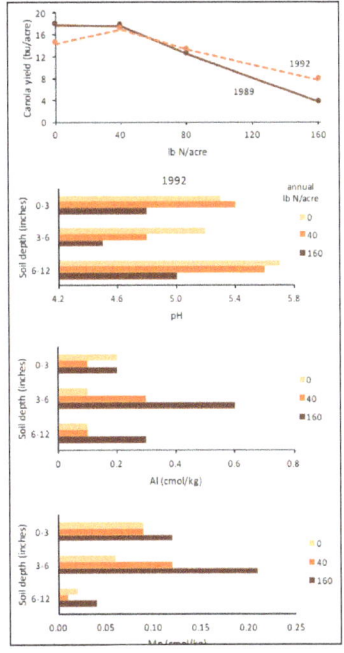

Annual applications of excess urea for 9 years (1983 to 1992)
decreased canola yield and soil pH, and increased aluminum and manganese to
potentially toxic levels. The suggested nitrogen rate was 40 lb N/acre

Many commercial liming products are a mixture of particle sizes to provide both a rapid increase in pH and continued neutralizing action over a few years' time.

Fertilizer

Nitrate-based nitrogen fertilizers, such as calcium nitrate (15.5-0-0 +19% Ca) may increase soil pH at both the surface and deeper levels but only if the nitrate gets taken up by the plant and is not lost to leaching. In contrast, ammoniumbased fertilizers, such as urea (46-0-0) and ammonium phosphates (11-52-0 or 18-46- 0) can slowly lower pH of basic soils, yet in some areas of Montana and the inland Pacific Northwest have led to excessive acidification of the seeding zone and decreased yields. Soils in northern Montana declined from soil pH 6.2 to 5.6 in 10 years and in Idaho and eastern Washington from near neutral to pH less than 6 in about 25 years.

Nitrogen fertilizer acidification, and concomitant H^+, aluminum and manganese toxicity, is more severe with nitrogen application rates in excess of crop requirement, especially in the seeding zone. Considering the cost of lime to offset soil acidification induced by nitrogen fertilizer, the economically optimal nitrogen level was as much as 11.3 percent lower than the yield maximizing level when lime cost was considered. To minimize soil acidification due to nitrogen fertilizer, use practices that prevent excess nitrogen application, encourage uptake of all applied nitrogen, and reduce nitrate leaching.

Steps to minimize soil acidification due to fertilizer nitrogen

- Increase efficiency of nitrogen use.
  - Base nitrogen rate on spring soil test and realistic yeild potentiala.
  - Split nitrogen applications.
- Reduce nitrate loss.
  - Use slow-release nitrogen sources.
  - Use nitrogen sources with nitrification inhibitors.
  - Plant deep rooted crops to 'catch' deep nitrate.
- Consider non-ammonium based nitrogen sources.
  - Legume rotations.
  - Calcium ammonium nitrate (27-0-0).

Organic Material

Soil organic matter is the combination of plant and animal residues at various stages of decomposition and cells and tissues of soil organisms. The consistent benefit of SOM is that it buffers soil pH change. Soil organic matter offers many negatively charged sites to bind H^+ in an acidic soil, or from which to release H^+ in a basic soil, in both cases pushing soil solution towards neutral. Whether SOM changes soil pH in the long term depends on many factors.

When organic matter first begins to decay, it releases anions and cations. Plant foliage and stems generally contain more anions, so the initial decay over the first few weeks causes a soil pH increase. This initial increase in soil pH, especially from high nitrogen plant residue, could be used to reduce H^+, aluminum or manganese toxicity in the seedling rooting zone long enough for seedling establishment. Soil microbes further break down the plant material to ammonium (mineralization) which temporarily increases pH. The ammonium gets converted to nitrate (nitrification) which causes pH to instead go down. If the nitrate is lost to leaching, pH drops even more. In the very long term, microbial decomposition decreases pH.

The net effect of organic matter addition on soil pH depends on the rate at which all these processes occur and what happens with the nitrogen produced (e.g., nitrate plant uptake vs. leaching loss), the quality and quantity of plant material, and initial soil pH. Soil pH will likely increase with decomposition of plants growing on basic soils, and manure derived from such plants, deep rooted plants that draw anions from deep soil layers to the soil surface, and, plant residue high in nitrogen.

Tillage

Tillage does not consistently increase or decrease soil pH. The top few inches of no-till soils can become more acidic due to nitrogen fertilization in that zone. Occasional tillage mixes the acidic layer with higher pH sub-surface layers, or helps integrate lime treatment. A soil with 5 percent calcium carbonate, typical in Montana, contains around 100 tons of calcium carbonate in the top foot. This is enough to offset at least a century worth of acid forming nitrogen fertilizer if it could be tilled up into the acidic zone. However, tillage reduces SOM, therefore the soil's ability to resist change in soil pH.

Crop Selection

Crops vary in their ability to raise or lower soil pH. For example, harvest of high yielding leafy crops such as forage or corn can reduce soil pH because leaves and stems contain large amounts of base-forming cations (Ca^{2+}, K^+, Mg^{2+}). A grain harvest with plant residue left behind removes much smaller amounts of these elements.

Table: The processes and conditions that influence whether organic matter increases or decreases soil pH.

Increase pH	Decrease pH
Microbial decomposition of carbohydrates	
Mineralization to ammonium	Nitrification to nitrate
Volatilization loss of ammonia gas	Leaching loss of nitrate
High plant residue base-forming cation content	Low plant residue base-forming cation content
Large amount of residue	Small amount of residue
Soil pH < SOM pH	Soil pH > SOM pH

For example, oat straw requires 6 times the lime to counter the acidifying effect of its removal, than oat grain harvest, and alfalfa harvest requires 10 times the lime as oat grain harvest. However, removing residue is not a desirable practice to lower soil pH. The benefit of SOM from crop residue outweighs the potential soil acidification by residue removal.

Legumes acidify their rooting zone through nitrogen-fixation. The acidifying potential of annual legumes (pea<lentil<chickpea) is lower than that of perennial legumes.

Planting deep-rooted crops (e.g., safflower, sunflower, and winter wheat) helps prevent nitrate from leaching, thereby reducing soil acidification. Deep rooted crops can also pull base-forming cations from the subsurface to the surface. Acid and aluminum tolerant crops can be used to minimize nitrate leaching and add biomass to slow acidification while waiting for lime treatment to take effect

Testing Soil pH

Soil pH is measured to assess potential nutrient deficiencies, crop suitability, pH amendment needs, and to determine proper testing methods for other soil nutrients, such as phosphorus. Handheld pH sampling meters are now available that provide quick, reliable results from soil cores to determine soil pH at 1-inch or less increments. The process is not difficult, but the equipment does need regular cleaning, calibration and proper storage. Field testing with meters or 'color' kits can indicate whether alkalinity or acidity may be an issue and help select which soils to send to a laboratory. Field tests do not provide enough information to determine lime or sulfur requirements; laboratory buffer tests are necessary for lime or sulfate rate calculations.

Buffer tests tend to be regionally specific to account for a region's unique soil conditions. The Woodruff, SMS, Sikora, Mehlich or modified Mehlich tests are suitable for Montana soils (27). It is important to be aware of pH meters and buffer tests used and be consistent to ensure comparable data over time. Soil testing laboratories usually note test methods used on the soil test report. Also, pH varies seasonally, for example, a soil under wheat varied from pH 6.2 in early April, to 6.5 in midJune, to 5.5 in mid-October. Annual comparisons should be made from samples taken the same time of year.

Soils sampled for laboratory pH analysis should be 1 foot deep and divided into 0 to 3, 3 to 6, 6 to 9, and 9 to 12-inch depth increments. It is important to properly sample incremental depths because a given pH zone can exist in only a narrow depth increment, for example, the top 3 inches due to surface broadcast nitrogen fertilizer. Sampling over a 6 or 12-inch depth could seriously underestimate a soil pH decline in the critical seeding zone. Sampling only the top 3 inches would not allow one to determine if and how deep to plow to mix deep, higher pH soils with low pH surface soils.

Soil cores should be at least ¾-inch diameter and a composite of 6 to10 subsamples should be mixed and subsampled before sending in about a 2-fist size sample. Remove plant residue or duff on the soil surface before taking the soil sample core. Samples should be kept cold or frozen until delivered. Detailed soil sampling methods and laboratory selection are described in Soil Sampling and Laboratory Selection.

Soil Organic Matter

Organic Matter Cycling

Though living organisms are not considered within the technical definition of SOM, their presence is critical to the formation of SOM. Plant roots and fauna (e.g., rodents, earthworms, mites, and microorganisms) all contribute to the movement and breakdown of organic material in the soil.

As organic residues decompose, organic carbon and nutrients are either released for plant uptake or transferred to a more stable SOM pool. This process produces carbon dioxide through microbial respiration and chemical oxidation, which is eventually released to the air.

The three main pools of SOM, determined by their time for complete decomposition, are dissolved organic matter (DOM, 1-2 years), particulate organic matter (POM, 15-100 years) and humus (500-5,000 years). Both DOM and POM are biologically active, meaning they are continually being decomposed by microorganisms, thereby releasing many organically-bound nutrients, such as nitrogen, phosphorus, and other essential nutrients, back to the soil solution. Dissolved organic matter is primarily composed of soluble portions of fresh plant and animal residues. That which is not completely decomposed moves into the POM pool, consisting primarily of detritus (cells and tissues of decomposed material). Particulate organic matter is partially resistant to microbial decomposition and serves as an important long-term supply of nutrients.

In contrast to DOM and POM, humus is not biologically active and is the pool responsible for many of the soil chemical and physical properties associated with SOM and soil quality. Representing approximately 35 to 50 percent of total SOM, humus is a dark, complex mixture of organic substances modified from original organic tissue, synthesized by various soil organisms, and resistant to further microbial decomposition. Because of this, humus breaks down very slowly and may exist in soil for hundreds or even thousands of years. Due to its chemical make-up and reactivity, humus is a large contributor to a soil's ability to retain nutrients on exchange sites. Humus also supplies organic chemicals to the soil solution that can serve as chelates and increase metal availability to plants. Additionally, organic chemicals have been shown to minimize the binding of phosphate with calcium (creating an insoluble mineral), possibly keeping fertilizer phosphorus in soluble form for a longer period. Dissolved organic chemicals act to 'glue' soil particles together, enhancing aggregation and increasing overall soil aeration, water infiltration and retention, and resistance to erosion and crusting. Soils high in humus are dark brown or black, increasing the amount of solar radiation absorbed by the soil and thus, soil temperature.

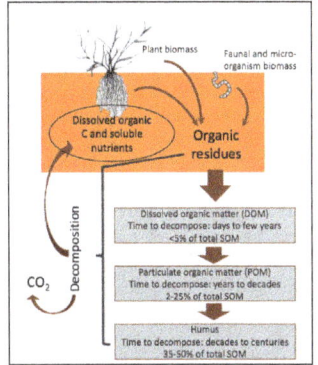

Organic matter decomposition cycle

SOM Decomposition and Accumulation

Soil organic matter content depends on the balance between organic residue addition and the rate of decomposition. Good growing conditions, crops or cover crops rather than fallow, and leaving plant residue on the field all add organic matter. Soil organic matter decomposition rates depend on SOM form, soil texture and drainage, carbon:nitrogen ratios of organic materials, climate, and cropping practices. Neither excess buildup (peat), nor rapid decomposition are ideal. Once a portion of SOM decomposes, its benefits to soil aeration, and nutrient and water holding capacity are decreased, and carbon is lost from the system as carbon dioxide gas.

As previously noted, SOM forms (i.e., DOM, POM, or humus) accumulate and decompose at different rates. For example, DOM and POM levels can fluctuate relatively quickly with changes in land management practices, particularly the adoption of no-till systems and recropping. Research has shown DOM and POM levels to increase in no-till systems compared to conventional till systems (33, 30), yet levels may decline following a return to tillage or under certain climatic conditions (discussed below). Humus content, on the other hand, is much more constant. Since SOM tests do not differentiate between SOM forms, changing DOM and POM levels can cause fluctuations to occur in total SOM levels, even though humus content remains the same.

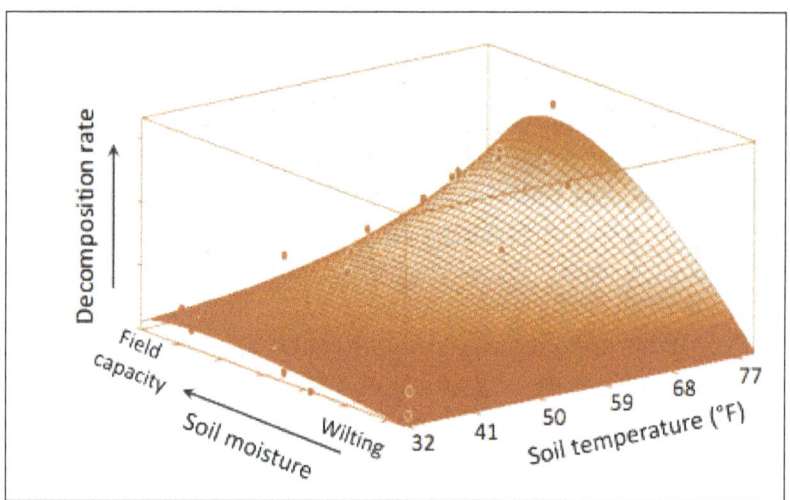

SOM decomposition increases as the combination of
temperature and soil moisture increases within conditions favorable
for most microbial growth. The dots are measured values (34, Massachusetts)

Soils high in clay and silt are generally higher in SOM content than sandy soils. This is attributed to restricted aeration in finertextured soils, reducing the rate of organic matter oxidation, and the binding of humus to clay particles, further protecting it from decomposition. Additionally, plant growth is usually greater in fine-textured soils, resulting in a larger return of residues to the soil.

Poorly-drained soils typically accumulate higher levels of SOM than well-drained soils. This is due to poor aeration causing a decline in soil oxygen concentrations. Many soil microorganisms involved in decomposition are aerobic (oxygen-requiring) and will not function well under anaerobic conditions (oxygenlimiting). This anaerobic effect is evident in wetland areas in which the 'soil' is often completely composed of organic material.

The carbon:nitrogen ratio of organic material affects microorganism activity and subsequent decomposition rates. Organic materials with carbon:nitrogen ratios greater than 30:1 (e.g., cereal grain straw at 80:1) generally decompose slowly and tend to accumulate, whereas those with carbon:nitrogen less than 24:1 (e.g., pea cover crop) decompose quickly. To obtain a desired balance between SOM decomposition and accumulation, different crops can be planted in rotation or organic materials can be mixed.

Climate impacts decomposition and accumulation by affecting growth conditions for soil microorganisms. A combination of warm and moist soil is ideal for decomposition and rapid release of nutrients. Decomposition can be faster at higher temperatures, but adequate soil moisture becomes critical. At very low temperatures, decomposition is limited by both temperature and water availability, such as in cold deserts, and in arid and semi-arid portions of the northern Great Plains.

Cultivated land generally contains lower levels of SOM than comparable lands under natural vegetation. Prairie soils of the northern Great Plains originally had at least 4 percent SOM, whereas present day SOM content in most Montana and Wyoming agricultural topsoil ranges from 1.5 to 4 percent. In cultivated areas, only plant material remaining after harvest and not burned makes it back to the soil. Increasing crop residue by reducing fallow and appropriate fertilization helps increase SOM. In contrast, tillage reduces SOM in the plow layer because: a) it aerates the soil and breaks up organic residues, thus stimulating microbial activity and increasing SOM decomposition, b) decreases soil water, which decreases production, and, c) increases susceptibility to soil wind and water erosion. Minimizing tillage helps build SOM in the surface soil of our region, but planting cover crops with abundant fibrous roots (e.g., grass) instead of fallow, will likely do more to increase deep SOM than reducing tillage.

Chelation

As introduced in Micronutrients: Cycling, Testing, and Fertilizer Recommendations, many organic substances can serve as chelates for micronutrient metals. Chelates (meaning 'claw') are soluble organic compounds that bind metals such as copper, iron, manganese, and zinc, and increase their solubility and availability to plants. The dynamics of chelation are illustrated in figure. A primary role of chelates is to keep metal cations in solution so they can diffuse through the soil to the root. This is accomplished by the chelate forming a 'ring' around the metal cation that protects the metal from reacting with other inorganic compounds. Upon reaching the plant root, the metal cation either 'unhooks' itself from the chelate and diffuses into the root membrane or the entire metal-chelate complex is absorbed into the root, and then breaks apart, releasing the metal. Both cases can result in the metal being taken up by the root and the chelate returning to the soil solution to bind other metals.

Chelation may be particularly important for regions with basic soils. As previously noted, metal availability is often inhibited under basic soil conditions, causing plant micronutrient deficiencies. Iron, for instance, becomes nearly insoluble as soil pH nears 8 and chelation can greatly increase availability (up to 100-fold). Chelation can be increased through the use of commercial chelating agents, synthetic organic compounds such as ethylenediaminetetraacetic acid (EDTA), or by maintaining and increasing SOM levels.

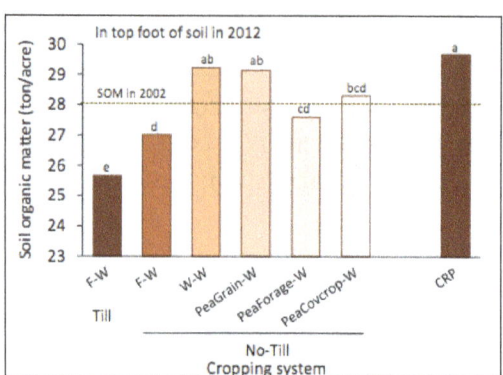

SOM after 10 years of cropping and tillage treatment. F - fallow, W - wheat, Pea harvested
for grain, as forage or killed early bloom as cover crop, CRP - Conservation Reserve Program = alfalfagrass.
Bars that have none of the same letters are different with 95% confidence

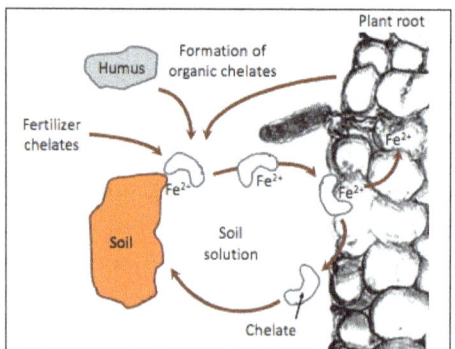

Cycling of chelated iron (Fe^{2+}) in soils

Carbon Sequestration

Carbon cycling is the transfer of both organic and inorganic carbon between the pools of the atmosphere (carbon dioxide and methane), terrestrial and aquatic organisms (living plants, animals, microorganisms), and the soil. Research within the last few decades indicates carbon concentrations in the atmosphere have increased with inputs linked to industrial emissions (i.e., extraction and combustion of fossil fuels) and land use changes (e.g., cutting and burning large areas of forest). This increase is causing the carbon balance between pools to shift and is contributing to global climate change. In response, the United States Department of Agriculture (USDA) along with other national and international organizations is promoting management practices to conserve and sequester (store) carbon. The goal of carbon sequestration is to reduce atmospheric carbon concentrations by taking carbon dioxide out of the atmosphere and storing it in 'sinks' such as soil. Increased carbon sequestration and changes in soil water dynamics following the reduction of fallow in the northern Great Plains since the 1970s coincides with a reduction in summertime temperatures across parts of this region (38, Canadian Prairies).

An important sink within soil is SOM, in which organic carbon (vs. inorganic, such as bicarbonate) levels are over twice as large as the atmosphere carbon dioxide pool and 4.5 times larger than the carbon pool in land plants (39). Soil carbon sequestration is accomplished through soil conservation practices that not only reduce soil erosion, but also increase the SOM content of soils. Possible conservation strategies which sequester carbon include converting marginal crop lands to perennial native systems (i.e., wildlife habitat) or rangelands, practicing notill or conservation-till farming, reducing the frequency of summer fallow in crop rotation, and incorporating, rather than disposing of organic amendments such as manure.

Producers can use the Natural Resources Conservation Service's (NRCS) COMNET-VR online tool to input management practices (e.g., residue management, cropping sequence, and tillage system) to estimate changes in soil carbon sequestration over time. The NRCS provides free technical assistance to develop and evaluate management practices.

SOM Testing

Soil organic matter levels are used to calculate nitrogen fertilizer rates, estimate a soil's water holding capacity and nutrient availability, and determine effects of agronomic practices on SOM over time. To sample for SOM, core the top 6 inches of soil. Organic material on the surface (i.e., duff or visible plant parts) should be excluded, as it is not part of SOM and can produce invalid readings. Soil testing laboratories return results as a SOM percentage. In interpreting SOM tests, it is important to understand what is being tested for and the test method used. Most SOM values are derived from organic carbon which represents approximately 50 percent of SOM, so a conversion factor of 1.7 to 2 is often used to estimate SOM concentrations (e.g., SOM = 1.7 x organic carbon). Common methods for testing SOM are Walkley-Black acid digestion and weight loss on ignition (LOI). Both of these methods test for total SOM and do not distinguish between different SOM forms, e.g., DOM, POM or humus. Therefore, two soils may have similar SOM contents, yet SOM function may differ considerably between the two soils. For example, one soil may have high humus, thus high mineral nutrient availability, yet be slow in releasing nitrogen, whereas a soil high in DOM could provide high nitrogen (e.g., from legume residue) but supply minimal micronutrients. Laboratory tests for SOM are not highly precise or reliable (41). For meaningful comparison of SOM over time or space, test results should come from the same laboratory.

Soil Acidity

An acid is a substance that tends to give up protons (hydrogen ions) to some other substance. Conversely, a base in any substance that tends to accept protons (hydrogen ions). Soil acidity may be defined as the soil system's proton (H^+ ions) donating capacity during its transition from a given state to a reference state.

Soil acidity involves intensity and quantity aspects. The intensity aspect is universally characterised by the measurements of H^+ ion activity, expressed as pH. The quantity aspect is characterised, directly or indirectly, by the quantity of alkali required to titrate soil to some arbitrarily established endpoint. Soil acidity is a major problem in relation to plant growth and therefore, acid soils are called a problem soil.

Source of Soil Acidity

Acid soil is a base unsaturated soil which has got enough of adsorbed exchangeable H+ ions so that to give soil a pH of lower than 7.0. The following important sources which are responsible for the development of acidic soils.

1. Leaching due to Heavy Rainfall

    Generally acid soils are common in all regions where rainfall or precipitation is high enough to leach appreciable amounts of exchangeable bases from the surface soils and relatively insoluble compounds of Al and Fe remains in soil. The nature of these compounds are acidic and its oxides and hydroxides react with water (H_2O) and release hydrogen (H^+) ions in soil solution and soil becomes acidic.

    Besides, when the soluble bases are lost, the H^+ ions of the carbonic acid and other acids developed in the soil replace the basic cations of the colloidal complex. As the soil gets gradually depletes of its exchangeable bases through constant leaching, it gets de-saturated and becomes increasingly acid.

    $$CO_2 H_2 O = HCO_3$$
    $$HCO_3 + CaCO_3 = Ca(HCO_3)_2 \downarrow$$
    $$\text{leachable}$$

2. Acidic Parent Material

    Some soils have developed from parent materials which are acid, such as granite and that may contribute to some extent soil acidity.

3. Acid Forming Fertilizers and Soluble Salts

    The use of ammonium sulphate $(NH_4)_2 SO_4$ and ammonium nitrate, $NH_4 NO_3$ increases soil acidity. Ammonium (NH_4) ions from $(NH_4)_2 SO_4$ when applied to the soil replace calcium (Ca^{2+}) ions from the exchange complex and the calcium sulphate ($CaSO_4$) is formed and finally leached out.

    $$(NH_4)_2 SO_4 \rightleftharpoons 2NH_4^+ + SO_4^{2-}$$

    $$2NH_4^+ + Ca\boxed{Caly} \rightarrow CaSO_4 + \begin{smallmatrix}NH_4\\NH_4\end{smallmatrix}\boxed{Caly}$$
    $$\downarrow$$
    $$\text{Leached out}$$

    $$\begin{smallmatrix}NH_4\\NH_4\end{smallmatrix}\boxed{Caly} + 3O_2 \xrightarrow{\text{Nitrification}} \begin{smallmatrix}H\\H\end{smallmatrix}\boxed{Caly} + 2HNO_3$$
    $$\text{Acid soil}$$

Besides, basic portion of ammonium sulphate, $(NH_4)_2 SO_4$ is NH_4 and it undergoes biological transformation in the soil and form acid forming nitrate (NO_3^-) ions. Similarly, sulphur also produce acid forming sulphate (SO_4^{2-}) ions through oxidation.

Divalent cations of soluble salts usually have a greater effect on lowering soil pH than monovalent metal cations.

4. Humus and Other Organic Acids

    Humus materials in soils occur as a result of microbiological decomposition of organic matter and contain different functional groups like carboxylic (-COOH), phenolic (-OH) etc. which are capable of attracting and dissociating hydrogen (H^+) ions.

During organic matter decomposition, humus, organic acids and different acid salts may also be produced and also concentration of CO_2 increased. The increased concentration of CO_2, hydrolysis of acid salts and various organic acids combinedly increased the total acidity of soil.

5. Aluminosilicate Minerals

    At low pH values most of the aluminium (Al) is present as the hydrated aluminium ions (Al^{3+}) which undergoes hydrolysis and release hydrogen (H^+) ions in the soil solution.

$$Al^{3+} + H_2O \xrightleftharpoons{\text{Hydrolysis}} Al(OH)^{2+} + H^+$$

$$Al(OH)^{2+} + H_2O \xrightleftharpoons{} Al(OH)_2^{+} + H^+$$

It is also possible that structural OH^- ions at corners and edges may dissociate hydrogen (H^+) ions and develop soil acidity.

6. Carbon Dioxide (CO_2)

    Soil containing high concentration of CO_2, the pH value of such soil will be low i.e. the soil becomes acidic. Root activity and metabolism may also serve as sources of CO_2 which ultimately helps the soil to become acidic.

7. Hydrous Oxides

    These are mainly oxides of iron and aluminium. Under favourable conditions they undergo stepwise hydrolysis with the release of hydrogen (H^+) ions in the soil solution and develop soil acidity.

8. Aluminium and Iron Polymers

    The Al^{3+} ions displaced from clay minerals by cations are hydrolyzed to monomelic and polymeric hydroxy aluminium complexes. Hydrolysis of the monomelichexaquoaluminium or iron forms are illustrated by the following stepwise reactions with the liberation of hydronium ion (H_3O^+) and lower soil pH. Each successive step occurs at a higher pH.

$$Al(H_2O)_6^{3+} + H_2O \rightleftharpoons Al(OH)(H_2O)_5^{2+} + H_3O^+$$

$$Al(OH)(H_2O)_5^{2+} + H_2O \rightleftharpoons Al(OH)_2(H_2O)_4^{+} + H_3O^+$$

$$Al(OH)_2(H_2O)_4^{+} + H_2O \rightleftharpoons Al(OH)_3(H_2O)_3^{0} + H_3O^+$$

$$Al(OH)_3(H_2O)_3^{0} + H_2O \rightleftharpoons Al(OH)_4(H_2O)_2^{-} + H_3O^+$$

Similary. $Fe(H_2O)_6^{3+} + H_2O \rightleftharpoons Fe(OH)(H_2O)_5^{2+} + H_3O^+$
(reaction occurs under more acidic conditions than Al)

$$Fe(OH)_3(H_2O)_3^{0} + H_2O \rightleftharpoons Fe(OH)_4(H_2O)_2^{-} + H_2O^+$$

Kinds of Soil Acidity

Broadly soil acidity may be of two kind's viz:

  i) Active acidity and

  ii) Potential reserve/ exchange acidity.

The nature of soil can be illustrated as follows:

$$\text{Adsorbed H (and AI) ions on soil colloids} \rightleftarrows \text{Soil solution H (and AI) ions}$$
$$\text{(potential/exchange/reserve acidity)} \qquad\qquad\qquad\qquad \text{(active acidit)}$$

  i) Active Acidity:

  Active acidity may be defined as the acidity develops due to hydrogen (H^+) and aluminium ($Al)^{3+}$ ions concentration of the soil solution. The magnitude of this acidity is limited.

  ii) Exchange Acidity:

  Exchange acidity may be defined as the acidity develops due to adsorbed hydrogen (H^+) and aluminium (Al^{3+}) ions on the soil colloids. The magnitude of this exchange acidity is very high. However, residual acidity may be included to the total acidity.

  Residual acidity may be defined as the acidity which remains in the soil after active and exchange acidity has been neutralized. It is associated with aluminium-hydroxy ions and with H and Al atoms that are bound in non-exchangeable forms by organic matter and silicate clay.

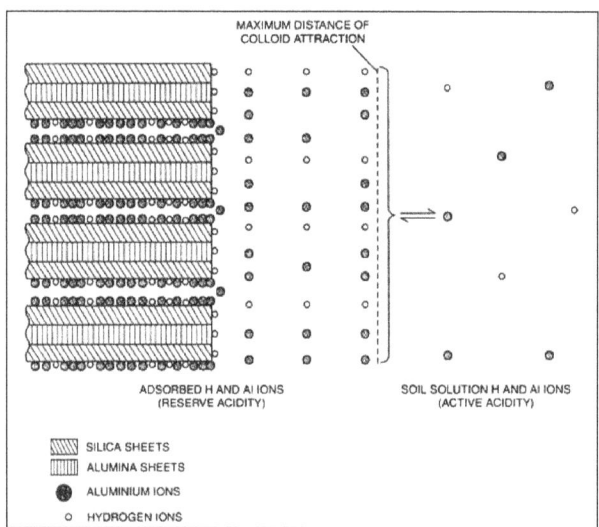

Relationship between reserve or exchange and active acidity

Total Acidity

The total acidity is summation of active, exchange and residual acidity. It can be written as,

Total acidity = Active acidity + Exchange acidity + Residual acidity.

Therefore, total soil acidity depends on the active, exchange and residual acidity of the soil.

Active acidity – Soluble acidity, in the solution.

Reserve acidity – Adsorbed acidity, on the surface of particles.

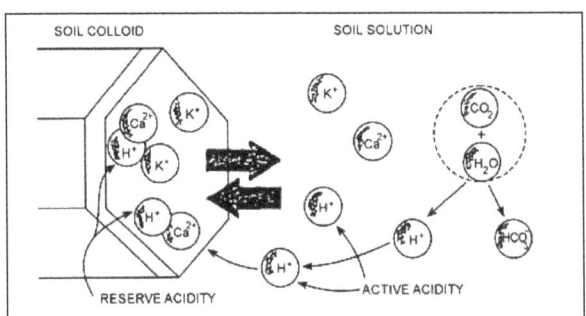

Adsorbed acidic ions : H^+, Al^{3+}, Fe^{3+}

Why are the trivalent ions acidic?

$$\left[Al(H_2O)_6\right]^{3+} = \left[Al(OH)(H_2O)_5\right]^{2+} + H^+$$

$$\left[Al(OH)(H_2O)_5\right]^{2+} = \left[Al(OH)_2(H_2O)_4\right]^{+} + H^+$$

$$\left[Al(OH)_2(H_2O)_4\right]^{+} = \left[Al(OH)_3(H_2O)_3\right] + H^+$$

$$\left[Al(OH)_3(H_2O)_3\right] = \left[Al(OH)_4(H_2O)_2\right]^{-} + H^+$$

$$\left[Al(OH)_4(H_2O)_2\right]^{-} = \left[Al(OH)_5(H_2O)\right]^{2-} + H^+$$

$$\left[Al(OH)_5(H_2O)\right]^{2-} = \left[Al(OH)_6\right]^{3-} + H^+$$

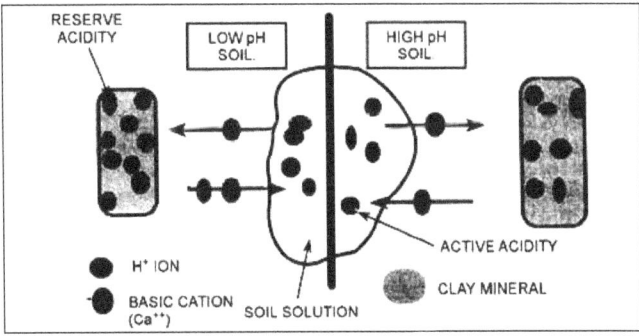

Diagram of soil Ph and active and reserve acidity

Problems of Soil Acidity

Problems of soil acidity may be divided into three groups which are as follows:

1. Toxic effects

    a) Acid toxicity: The higher hydrogen ion concentration is toxic to plants under strong acid conditions of soil. The acid toxicity includes possible toxicities of acid anions as well as H^+ ions.

    b) Toxicity of different nutrient elements

2. Nutrient availability

    a) Non-specific effects

    b) Specific effects

       ○   Exchangeable bases

       ○   Nutrient imbalances

3. Microbial activity

Iron and Manganese

The concentration of these two ions (Fe^{2+} and Mn^{2+}) in soil solution depends upon the soil reaction or pH, organic matter and intensity of soil reduction. Due to increase in organic matter content in the soil the population of soil microbes increases and very rapidly used up the soil oxygen and results reduction of soil.

As a result of soil reduction, the nutrient elements like Mn^{4+} and Fe^{3+} reduce to Mn^{2+}(manganous manganese) and Fe^{2+} (ferrous iron) respectively and increases their concentration to a very high and toxicity of those elements develops. Due to such toxic effects, a physiological disease of rice is found in submerged soils which are popularly known as browning disease.

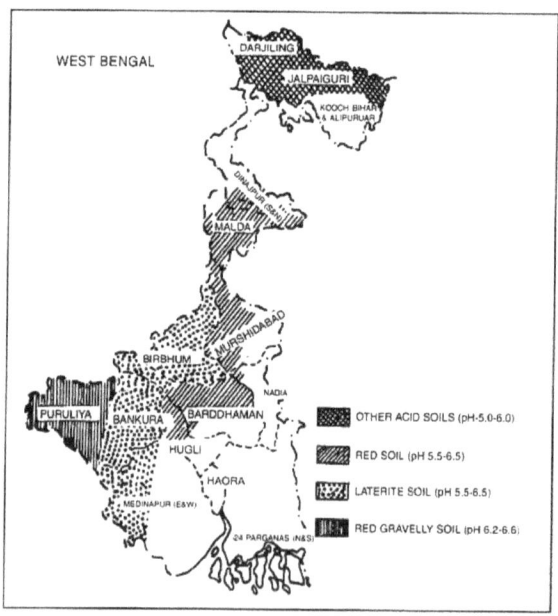

Acid soil map of map of West Bengal

Toxicity of Aluminium (Al)

The toxicity of aluminium may be greatly influenced by the accompanying cations. The toxicity of aluminium tends to decrease with an increase in the concentration of other cations such as calcium. Aluminium toxicity is a problem in both upland and lowland soils.

Aluminium Toxicity in Soils Affects Plant Growth in Various Ways

1. It restricts the root growth.

2. It affects various plant physiological processes like division of cells, formation of DNA and respiration etc.

3. It restricts the absorption and translocation of some important nutrient elements from soil to the plant like phosphorus, calcium, iron, manganese etc.

4. It causes wilting of plants.

5. It also inhibits the microbial activity in the soil.

Nutrient Availability

a. Non-specific effects

    It is associated with the inhibition effect of root growth and thereby affects the nutrient availability.

b. Specific Effects.

Exchangeable Bases

There are two aspects of availability of exchangeable bases i.e. ion uptake process and the release of bases from the exchangeable form may be adversely affected due to soil acidity. Due to complementary ion effect exchangeable bases are released preferentially in a fractional exchange. Deficiency of bases like Ca^{2+} and Mg^{2+} are found in acid soils.

Nutrient Imbalances

It is evident that soluble iron, aluminium and manganese are usually present in their higher concentrations under moderate to strong acid soils. Phosphorus reacts with these ions and produces insoluble phosphatic compounds rendering phosphorus unavailable to plants. Besides these, fixation of phosphorus by hydrous oxides of iron and aluminium or by adsorption, the availability of phosphorus is decreased.

In acid soils, iron, manganese, copper and zinc are abundant, but molybdenum is very limited and unavailable to plants. In acid soils having very low pH, the availability of boron may also be decreased due to adsorption on sesquioxides, iron and aluminium hydroxy compounds. Nitrogen, potassium and sulphur become less available in an acid soil having pH less than 5.5.

Microbial Activity

It is well-known that soil organisms are influenced by fluctuations in the soil reaction. Bacteria and actinomycetes function better in soils having moderate to high pH values. They cannot show their activity when the soil pH drops below 5.5. Nitrogen fixation in acid soils is greatly affected by lowering the activity of Azotobacter sp.

Besides these, soil acidity also inhibits the symbiotic nitrogen fixation by affecting the activity of Rhizobium sp. Fungi can grow well under very acid soils and caused various diseases like root rot of tobacco, blights of potato etc.

Amelioration of Soil Acidity

In general the fertility status of acid soils is very poor and under strongly to moderately acidic soils the plant growth and development affect to a great extent. The crops grown on such problematic soils do not give remunerative return rather it lowers down the yield to a great extent.

Because of the limited land resource it needs judicious management practices so that yield of different crops can be increased. So one of the most important and practically feasible management practices is the use of lime and liming materials to ameliorate the soil acidity.

The addition of lime raises the soil pH to some prescribed value. This value is usually in the range of pH to some prescribed value. This value is usually in the range of pH 6.0 to 7.0, since this is an easily attainable value within the optimum range of most crop plants.

Lime requirement of an acid soil may be defined as the amount of liming material that must be added to raise the pH to some prescribed value. This value is usually in the range of pH 6.0 to 7.0, since this is an easily attainable value within the optimum range of most crop plants.

Liming Reactions

Lime reactions in soils depend upon the nature and the fineness of the liming materials. Lime is usually applied to soils in the form of ground limestone. Limestone's can be classified as calcitic ($CaCO_3$), dolomite [$CaMg(CO_3)_2$] or a mixture of the two.

Both of these limestone's are sparingly soluble in pure water but do become soluble in water containing carbon dioxide. The greater the partial pressure of carbon dioxide in the system, the more soluble the limestone becomes. Dolomite is somewhat less soluble than calcite.

The Reaction of Limestone ($CaCO_3$) can be written as:

$$CaCO_3 + H_2O + CO_2 \rightarrow Ca(HCO_3)_2$$

$$Ca(HCO_3)_2 \rightarrow Ca^{2+} \downarrow + 2HCO_3^- \underset{\text{exchange reactions}}{\binom{\text{takes part in cation}}{}}$$

$$\underset{\binom{\text{from soil}}{\text{solution}}}{H^+} + \underset{\binom{\text{from}}{\text{lime}}}{HCO_3^-} \rightarrow H_2CO_3 \rightleftarrows H_2O + CO_2$$

In this way hydrogen ions (H^+) in the soil solution react to form weakly dissociated water, and the calcium (Ca^{2+}) ion from limestone is left to undergo cation exchange reactions. The acidity of the soil is, therefore, neutralized and the per cent base saturation of the colloidal material is increased.

The process of changing pH by the addition of lime [$Ca(OH)_2$] is illustrated in figure.

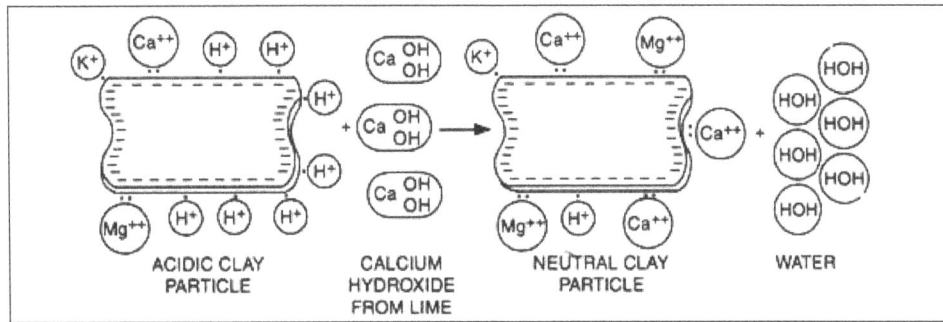

Change of pH with the application of time

Factors Affecting Liming Reactions

Various environmental factors affect the rate of limestone reaction. A few of the important factors are being discussed here:

1. Moisture

   The greater the amount of moisture, the more rapid is the rate of reaction. Obviously moisture must be present before the solubility reaction can occur. As moisture increases, the degree of aeration are reduced resulting an increase in the concentration of carbon dioxide (CO_2) in the soil air and thereby increases the rate of reaction. The increased moisture would also allow for a greater volume of solution and, therefore, a lower concentration of reaction end products. Since the reaction is an equilibrium reaction, the accumulation of end products would reduce reaction rate over time.

2. Temperature

   Lime and liming materials react more rapidly at high that at low temperatures. This effect is probably related to diffusion rates of end products away from the reaction sites.

3. Amount of exchange acidity

   The amount of exchange acidity present in the soil affects reaction rate. If a soil has a high lime requirement and if a sufficient quantity of limestone is added to neutralize the acidity present, the initial reaction will be quite rapid. However, as the acidity becomes neutralized, the rate of reaction decreases and finally, as neutrality is approached becomes almost negligible.

Liming Materials

Liming materials may be defined as materials that are necessary for the neutralization of soil solution hydrogen (H^+) ions. The materials commonly used for the liming of soils are the oxides, hydroxides, carbonates and silicates of calcium or calcium and magnesium.

The presence of only these elements does not consider a material as a liming compound. In addition to these compounds, the accompanying anion must be one that will reduce the activity of hydrogen (H^+) ions and hence aluminium in the soil solution. These are called "Agricultural liming materials".

Reason why Gypsum is not Considered as a Liming Material

Gypsum is not considered as liming materials because on its application to an acid soil it dissociates into calcium (Ca^{2+}) and sulphate (SO_4^{2-}) ions:

$$CaSO_4 \: D \: Ca^{2+} + SO_4^{2-}$$

The accompanying anion is sulphate and it reacts with soil moisture produces mineral acid (H_2SO_4) which also increases soil acidity instead of reducing soil acidity.

$$SO_4^{2-} + H_2O \rightarrow \underset{(\text{strong mineral acid})}{H_2SO_4}$$

Besides this, calcium (Ca^{2+}) in the gypsum after dissociation will result in replacement of adsorbed aluminium (Al^{3+}) in a localized soil zone (when gypsum applied as band placement) with a significant lowering of soil pH. Therefore, gypsum does not qualify as a liming material.

Kinds of Liming Materials

There are various kinds of liming materials that are used for the correction of soil acidity.

1. Oxides of lime

    It is normally called burned lime or quick lime. Oxide of lime is more caustic than limestones.

Burned lime is produced by heating limestone and dolomite as follows:

$$\underset{(\text{Limestone})}{Ca\,CO_3} + Heat \qquad \rightarrow CaO + CO_2 \uparrow$$

$$CaMg(CO_3)_2 + heat \rightarrow CaO + MgO + 2CO_2 \uparrow$$

2. Hydroxides of lime

    It can be produced by adding water to burned lime and is called slaked lime.

$$\underset{(\text{Burned lime})}{CaO} + H_2O \rightarrow \underset{(\text{Slaked lime})}{Ca(OH)_2}$$

    It is more caustic than burned lime (CaO).

If it is kept open in the moist air, then combination of calcium hydroxide occurs as follows:

$$Ca(OH)_2 + Co_2 \text{â †'} CaCO_3 + H_2O$$

In case of $Mg(OH)_2$

$$Mg(OH)_2 + CO_2 \text{â †'} MgCO_3 + H_2O$$

3. Carbonates of lime

    These are by products of certain industries and so the content of calcium and magnesium varies. The two important minerals are found in this group- Calcite ($CaCO_3$) and dolomite [$CaMg(CO_3)_2$].

4. Slags

    These are generally three types of slags that are found important:

a) Blast furnace slag: It is a by-product of the manufacture of pig iron. As a liming material, this slag behaves essentially as calcium silicate. The neutralizing value of blast furnace slags ranges from about 75-90%.

b) Basic slag: It is a by-product of the basic open-hearth method of making steel from pig iron, which in turn is produced from high phosphorus iron ores. The impurities in the iron, including silica and phosphorus are fluxed with lime and the basic slags are produced. Its neutralizing value ranges from 60-70%.

c) Electric furnace slag: This is produced from the electric furnace reduction of phosphate rock during preparation of elemental phosphorus. This product is largely calcium silicate and is used as a liming material.

5. Other liming materials

    Coral shell, chalk, wood ash, press mud, by-product material of paper mills, sugar factories, fly ash and sludge etc. are considered as liming materials and also used for the amelioration of soil acidity.

Efficiency of Liming Materials

The efficiency of liming materials can be judged on the basis of following important factors because of varying neutralizing capacity of different liming materials:

1. Neutralizing value (N.V.) or calcium carbonate equivalent of liming materials,

2. Purity of liming materials,

3. Degree of fineness of liming materials.

1. Neutralizing value (N.V.) or Calcium Carbonate Equivalent (CCE)

Calcium carbonate equivalent (CCE) is defined as the acid neutralizing capacity of an agricultural liming material expressed as a weight percentage of calcium carbonate.

$$\text{CCE of a liming material} = \frac{\text{Molecular weight of Ca CO}_3}{\text{Molecular weight of a liming material whose CCE is to be determined}} \times 100$$

Example:

Calculate the CCE of dolomite, $CaMg(CO_3)_2$.

Solution:

In dolomite, there is mixture of $CaCO_3$ and $MgCO_3$. Both are almost equally effective in neutralizing soil acidity. Considering the materials as two separate liming materials present in the dolomite.

So the molecular weight of dolomite may be considered as 184/2 = 92 since 84 gms of $MgCO_3$ will neutralize the same amount of acid as 100 gms of $CaCO_3$.

So,

$$\text{CCE of dolomite} = \frac{\text{Molecular weight of Ca CO}_3}{\text{Molecular weight of dolomite}} \times 100$$

= 100/92 Ã— 100 = 108.7

Neutralizing value or CCE of some liming materials is given in table.

Table: Neutralizing Value of Some Liming Materials

Liming materials'	Neutralizing value of CCE (%)
Calcium oxide(CaO)	179
Calcium hydroxide[Ca(OH)$_2$]	136
Dolomite [CaMg(Co$_3$)$_2$]	108.7
Calcite (CaCo$_3$)	100
Basic slag(CaSiO$_3$)	86

2. Purity of liming materials

The more purer the liming material, the higher will be its effectiveness for the amelioration of soil acidity.

3. Degree of fineness of liming materials

The rate of reaction of liming materials with an acid soil depends upon its fineness because finer materials increase the surface contact with the soil. If the liming materials are coarse, the reaction will be slight. The magnitude of such reaction as affected by the fineness of liming materials can be evaluated by measuring the change of soil pH. The amount of finer fraction of liming materials will be required much less as compared to coarser fractions of the material to achieve a certain pH.

The fineness is measured in terms of the ability of a material to pass through a sieve having 60 holes of equal size in one linear inch. Such a sieve is called a 60 mesh sieve and a material passing through such a sieve is allotted 100% efficiency rating.

One such scale used in Ohio (USA) is given below:

	Efficiency of rating (%)
1. Material passing through a 60 mesh sieve	100
2. Material passing through a 20 mesh but not a 60 mesh sieve	60
3. Material passing through an 8 mesh but not a 20 mesh sieve	20

Percent Effective Calcium Carbonate (ECC) or (Neutralizing Index)

The effective calcium carbonate (ECC) rating of a limestone or liming materials is one product of its calcium carbonate equivalent (CCE) and the fineness factor. The fineness factor is the sum of the product of the percentage of material in each of the three size fractions multiplied by the appropriate effectiveness factor.

Per cent ECC or N.I. = CCE Ã— fineness factor

Chemical Reactions between Liming Materials and Acid Soils during the Process of Amelioration:

1. Oxides of lime (CaO)

   When oxides of lime like CaO are applied to an acid soil, it reacts almost immediately as follows:

$$\text{Soil} \begin{bmatrix} H \\ H \\ H \\ AI \\ AI \end{bmatrix} + H_2O + 2CaO \rightarrow \begin{bmatrix} H \\ Ca \\ Ca \\ AI \\ H \end{bmatrix} \text{Soil} + AI(OH)_3$$

2. Hydroxides of lime Ca(OH)$_2$

   When hydroxides of lime like Ca(OH)$_2$ is applied for the reclamation of an acid soil, the following chemical reaction takes place:

$$\text{Soil} \begin{bmatrix} H \\ H \\ H \\ AI \end{bmatrix} + 2Ca(OH)_2 \rightarrow \begin{bmatrix} H \\ Ca \\ Ca \\ H \end{bmatrix} \text{Soil} + AIAI(OH)_3 + H_2O$$

3. Carbonates of lime (CaCO$_3$)

   When carbonates of lime like calcite is applied to an acid soil, a part of CaCO$_3$ undergoes solution and combines with H$_2$CO$_3$ to form soluble Ca(HCO$_3$)$_2$.

This calcium bicarbonate in solution form reacts with the soil colloids with the evolution of CO2 as follows:

$$CaCO_3 + H_2CO_3 = Ca(HCO_3)_2$$
$$Ca(HCO_3)_2 + {}_H^H[\text{Soil} = Ca[\text{Soil} + 2H_2O + CO_2$$

And the rest portion of the limestone goes in close contact with the soil colloids in solid condition as follows:

$${}_H[\text{Soil} + CaCO_3 = Ca[\text{Soil} + \quad H_2O + CO_2$$

Solid condition

4. Basic slag

Basic slag can also be used as liming material when it is applied to an acid soil the following chemical reaction takes place.

$$\text{Soil}\big]_{Al}^{H} + 2\,CaSiO_3 + 3\,H_2O = _{Ca}^{Ca}\big[\text{Soil} + \underset{(\text{Metasilicic acid})}{2\,H_2\,SiO_3} + Al(OH)_3$$

The metasilicic acid is weakly dissociated, much less so that the clay absorbed hydrogen (H⁺) ions and the pH of the soil is raised.

Besides these chemical amendments, some other management practices have been found to be effective for the amelioration of an acid soil namely:

i)   Use of basic fertilizers like sodium nitrate, basic slag etc.

ii) Reducing leaching of basic cations by following appropriate soil and water management practices.

iii) Acid tolerant crops like rice, potato, tea wheat, sweet potato, maize, brinjal etc. should be cultivated.

Lime Requirement and Liming Factor

Once a soil is found to be acidic and then it requires some management practices either chemical amendment with some liming materials or some cultural management practices in order to improve its fertility status through the modification of soil reaction to a favourable position.

It has been found that the correction of soil acidity as well as improvement of soil fertility through liming is most effective. The different intensity of soil acidity required different amount of lime. So determination of lime requirement from an acid soil is prerequisite in a liming programme.

The desirable soil pH range for most of the field crop is 6.0 to 7.0. The amount of lime required to be added to acidic soil to raise the pH of that soil to a desired value is known as lime requirement. Shoemaker et al. (1961) buffer method is used for the determination of lime requirement of an acid soil.

Buffer method for determining lime requirement of soils appreciable amount of extracted aluminium. (Proc. Soil Sci. Soc. Amer. 25: 274). Lime requirement in terms of pure calcium carbonate can be observed from the table.

Table: Lime Requirement of an Acid Soil

pH of soil buffer suspension (Field soil sample)	Lime required to bring pH down to indicated level ($CaCo_3$) in tonnes per acre		
	pH 6.0	pH 6.4	pH 6.8
6.7	1.0	1.1	
6.6	1.4	1.7	
6.5	1,8	1...5	
6.4	2.3	2.7	
6..3	2.7	3.7	

6.2	3.	3.7	
6.1	3,5	4,2	
6.0	4.7	5.4	
5.9	44	5.2	
5.8	4.8	5.7	
5.7	5	6.2	
56	5..6	1.7	
5,5	6.0	1_2	
5.4	6.5	8.9	
5.3	6.9	K.2	
5.2	7.4	8,4	
5.I	7,8	9.1	
5.0	8.2	9.6	
4.9	8.6	10.1.	
4.8	9.1	10.6	

Liming Factor

'Liming factor' may be defined as the factor by which the actual amount of lime can be calculated from the estimated theoretical amount of lime. This factor varies from 1 to 3, depending on rate of limestone solution, plant uptake and leaching during the reaction period. A liming factor of 1.5 to 2.0 is generally used when the theoretical amount of lime necessary to bring the soil to a given pH is converted to the actual amount of limestone to be added in the field.

Methods of Applying Lime

The application of small amounts of lime of soil in every year or twice in a year has been found to be most effective. But this involves the application cost of the liming material. Generally lime should be applied well ahead of the crop cultivation and broadcasted limes are to be well mixed with the whole plough layer soil so that liming reaction can occur in a faster rate.

The usual liming practice consists of a compromise between what is most effective and what is cheapest per ton of lime applied. When both surface and sub-surface soils are strongly acidic e.g. ultisols, it sometimes pays to incorporate lime to a depth of about 30 cm (12 inches). Application of lime of no till field soils may not be as effective as application to an equivalent cultivated soil.

Lime Balance Sheet

When a soil has had its acidity corrected by lime, how often must lime be added and how much is needed to keep the soil pH suitable? The answer depends upon the rate of lime loss.

Lime is neutralized or lost from the soil by six activities:

1. Neutralization by acid-forming fertilizers, a rapid change.

2. Neutralization by the acid formed by CO_2 in water (from organic matter decomposition), a slow process.

3. Leaching a relatively slow change.

4. Removal in harvested or grazed crops; relatively slow loss.

5. Erosion, a top soil is lost with its higher base saturation and often leaving more acidic subsoil to be limed.

6. Neutralization by acids dissolved in precipitation, resulting from oxides of sulphur fumes from manufacturing plants, a slow process.

Effect of over Liming

When excessively large amounts of lime applied to an acidic soil the growth of plants is affected by influencing either one or many of these following causes:

1. Deficiency of iron, copper and zinc will occur.

2. Phosphorus and potassium availability will be reduced.

3. Due to high OH^- ion concentration by over liming, root development will be inhibited in association with tip swelling brought about by hydrations. Due to dehydrating properties of boron, it acts as a protective agent for excess OH^- ion concentration.

4. Due to over liming, boron deficiency will occur.

$$2\,AIx_3 + 3\,CaCO_3 + 3\,H_2O = 3\,Cax_2 + \underset{(\text{precipitation})}{2\,AI(OH)_3 \downarrow} + 3CO_2$$

where 'x' is the exchange site.

The freshly $Al(OH)_3$ is then available for adsorption of boron. This reaction mechanism is involved for causing boron deficiency.

5. Due to application of lime in excess, the incidence of diseases like scab in root crops will be increased.

All these effects can be reduced with the application of large amounts of organic manures like well rotten farm yard manure, green manure crops, compost, phosphorus, boron or a mixture of micronutrient fertilizers to the soil.

Soil Salinity and Sodicity

Saline irrigation water contains dissolved substances known as salts. In much of the arid and semi-arid United States (including Montana), most of the salts present in irrigation water are chlorides, sulfates, carbonates, and bicarbonates of calcium magnesium, sodium, and potassium. While salinity can improve soil structure, it can also negatively affect plant growth and crop yields.

Sodicity refers specifically to the amount of sodium present in irrigation water. Irrigating with water that has excess amounts of sodium can adversely impact soil structure, making plant growth difficult. Highly saline and sodic water qualities can cause problems for irrigation, depending on the type and amount of salts present, the soil type being irrigated, the specific plant species and growth stage, and the amount of water able to pass through the root zone.

Effects of Salinity on Plant Growth

Salinity becomes a problem when enough salts accumulate in the root zone to negatively affect plant growth. Excess salts in the root zone hinder plant roots from withdrawing water from surrounding soil. This lowers the amount of water available to the plant, regardless of the amount of water actually in the root zone. For example, when plant growth is compared in two identical soils with the same moisture levels, one soil receiving salty water and the other receiving salt-free water, plants are able to use more water from the soil receiving salt-free water. Although the water is not held tighter to the soil in saline environments, the presence of salt in the water causes plants to exert more energy extracting water from the soil. The main point is that excess salinity in soil water can decrease plant available water and cause plant stress.

Soil water salinity is dependent on soil type, climate, water use and irrigation routines. For example, immediately after the soil is irrigated, plant available water is at its highest and soil water salinity is at its lowest. However, as plants use soil water, the remaining water is held tighter to the soil and becomes progressively more difficult for plants to obtain. As the water is taken up by plants through transpiration or lost to the atmosphere by evaporation, soil water salinity increases because salts become more concentrated in the remaining soil water. Thus, evapotranspiration (ET) between irrigation periods can further increase salinity. (Increased salinity due to ET is rarely taken into account in salinity charts.)

Effects of Salinity on Soil Physical Properties

Soil water salinity can affect soil physical properties by causing fine particles to bind together into aggregates. This process is known as flocculation and is beneficial in terms of soil aeration, root penetration, and root growth. Although increasing soil solution salinity has a positive effect on soil aggregation and stabilization, at high levels salinity can have negative and potentially lethal effects on plants. As a result, salinity cannot be increased to maintain soil structure without considering potential impacts on plant health.

Effects of Sodium and Sodicity on Soil Physical Properties

Sodium has the opposite effect of salinity on soils. The primary physical processes associated with high sodium concentrations are soil dispersion and clay platelet and aggregate swelling. The forces that bind clay particles together are disrupted when too many large sodium ions come between them. When this separation occurs, the clay particles expand, causing swelling and soil dispersion.

Soil dispersion causes clay particles to plug soil pores, resulting in reduced soil permeability. When soil is repeatedly wetted and dried and clay dispersion occurs, it then reforms and solidifies into almost cement-like soil with little or no structure. The three main problems caused by sodium-induced dispersion are reduced infiltration, reduced hydraulic conductivity, and surface crusting.

Salts that contribute to salinity, such as calcium and magnesium, do not have this effect because they are smaller and tend to cluster closer to clay particles. Calcium and magnesium will generally keep soil flocculated because they compete for the same spaces as sodium to bind to clay particles. Increased amounts of calcium and magnesium can reduce the amount of sodium-induced dispersion.

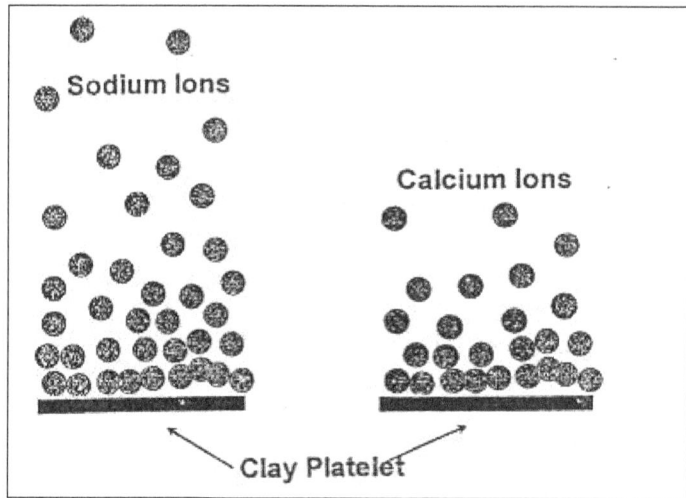

Behavior of sodium and calcium attached to clay particles

Infiltration

Soil dispersion hardens soil and blocks water infiltration, making it difficult for plants to establish and grow. The major implications associated with decreased infiltration due to sodium-induced dispersion include reduced plant available water and increased runoff and soil erosion.

Hydraulic Conductivity

Soil dispersion not only reduces the amount of water entering the soil, but also affects hydraulic conductivity of soil. Hydraulic conductivity refers to the rate at which water flows through soil. For instance, soils with well-defined structure will contain a large number of macropores, cracks, and fissures which allow for relatively rapid flow of water through the soil. When sodium-induced soil dispersion causes loss of soil structure, the hydraulic conductivity is also reduced. If water cannot pass through the soil, then the upper layer can become swollen and water logged. This results in anaerobic soils which can reduce or prevent plant growth and decrease organic matter decomposition rates. The decrease in decomposition causes soils to become infertile, black alkali soils.

Surface Crusting

Surface crusting is a characteristic of sodium affected soils. The primary causes of surface crusting are 1) physical dispersion caused by impact of raindrops or irrigation water, and 2) chemical dispersion, which depends on the ratio of salinity and sodicity of the applied water.

Surface crusting due to rainfall is greatly enhanced by sodium induced clay dispersion. When clay particles disperse within soil water, they plug macropores in surface soil by two means. First, they block avenues for water and roots to move through the soil. Second, they form a cement like surface layer when the soil dries. The hardened upper layer, or surface crust, restricts water infiltration and plant emergence.

Relationship between Salinity and Sodicity and Soil Physical Properties (EC/SAR)

The relationship between soil salinity and its flocculating effects and sodicity and its dispersive effects influence whether or not soil will stay aggregated or become dispersed under various salinity and sodicity combinations. As irrigation water with low salinity is applied to the soil by irrigation or rainfall, this water flows into the spaces between clay particles (micropores). If salinity of the applied water is low relative to soil salinity, swelling and dispersion of clay particles results. In contrast, irrigation water with higher salinity than the soil tends to cause particles to stay together, maintaining soil structure.

More than fifty years of research have been conducted to determine the relationship between salinity (EC) and sodicity (SAR) of irrigation water and its affects on soil physical properties. This relationship is now understood well enough to make accurate predictions of how specific soils will behave when irrigated water containing different levels of salts and sodium. The main concerns related to the relationship between salinity and sodicity of irrigation water are the effects on soil infiltration rates and hydraulic conductivities.

Swelling Factor

The ratio of salinity (EC) to sodicity (SAR) determines the effects of salts and sodium on soils. Salinity promotes soil flocculation and sodicity promotes soil dispersion. The combination of salinity and sodicity of soils is measured by the swelling factor, which is the amount a soil is likely to swell with different combinations of salinity and sodicity. Essentially, the swelling factor predicts whether sodium-induced dispersion or salinity-induced flocculation will more greatly affect soil physical properties.

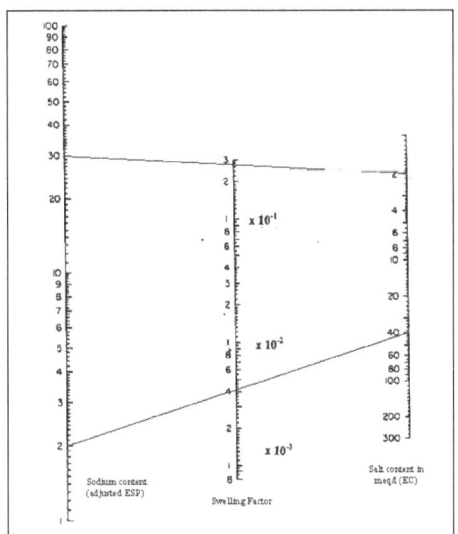

Swelling factor as a function of sodium content
(adjusted ESP) of soil and salt concentration of soil water

Scientists have been able to get a good idea of the swelling factor by using figure. It is possible to draw a line from the sodium content (adjusted ESP) in the left column to the appropriate salt concentration in the right column. The line intersects the middle column, the swelling factor, indicating how much the soil will swell. For instance, drawing a line between adjusted ESP = 2 and an

EC = 40 meq/L (red line) yields a swelling factor of 0.0041. In this example, the swelling factor of 0.0041 indicates that dispersion is not a problem. However, a combination of adjusted ESP = 30 and EC = 2 yields a swelling factor of 0.28 which indicates that dispersion is likely. In short, figure helps show how the dispersive effects of soils with high ESP can be lessened with the flocculating effects of irrigation water with high EC.

Infiltration Rates

Another approach to judging the effects of salinity (EC) and sodicity (SAR) on soil physical properties is to assess potential impacts of various irrigation water qualities on infiltration rates. Figure below demonstrates the relationship between salinity and sodicity and infiltration rates. For example, severe problems are likely if the irrigation water has low salinity and high sodicity. At SAR = 15, a severe reduction in infiltration will occur at an EC = 1 dS/m. An EC of 2.5 or less results in a slight to moderate reduction in infiltration. With an EC greater than 2.5, there will likely not be a reduction in infiltration. Similarly, numerically defines the relationship between EC, SAR, and infiltration rates.

Factors such as climate, soil type, crop and plant species and management practices also need to be accounted for when determining acceptable levels of salinity and sodicity of irrigation water. Rainfall also plays an important role in the relationship between salinity and sodicity and soil physical properties. Intense rainfall can flush salts beneath the root zone, but often cannot significantly reduce amounts of sodium bound to the soil. Therefore, rainfall can reduce the potential for soil aggregation from salts and increase the likelihood that sodium-induced dispersion will occur.

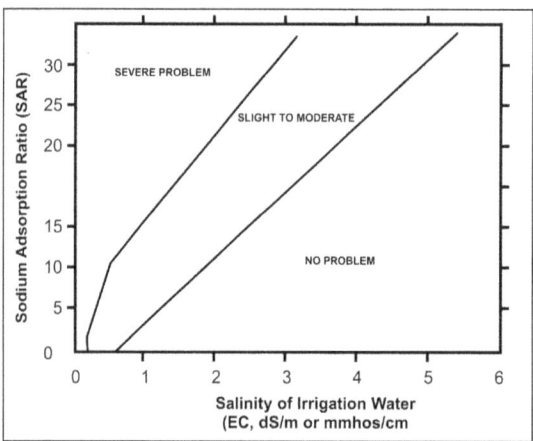

Potential for reduction in infiltration rates resulting from various
combinations of EC and SAR of applied water

SAR	EC dS/m No Problem	EC dS/m Slight to Moderate	EC dS/m Severe Problem
0 to 3	> 0.9	0.9 to 0.2	< 0.2
3 to 6	> 1.3	1.3 to 0.25	< 0.25
6 to 12	> 2.0	2.0 to 0.35	< 0.35
12 to 20	> 3.1	3.1 to 0.9	< 0.9
20+	> 5.6	5.6 to 1.8	< 1.8

Table: Guidelines for saline-sodic water quality suitable for irrigation, presented in terms of reduced infiltration.

Role of Soil Texture

Soil texture plays an important role in all aspects of irrigated agriculture, and the role of soil texture with respect to effects of salinity and sodicity is no exception. Soil texture helps determine how much water will be able to pass through the soil, how much water the soil can store, and the ability of sodium to bind to the soil.

Because they are composed of small particles, clay soils can hold more water and are slower to drain than course textured soils. Smaller particles can pack closely together, block the spaces between particles and prevent water from passing through. Sand particles are larger and therefore, have larger pore spaces for water to pass through. Under normal irrigation practices, sandy soils will naturally be able to flush more water through the root zone than clay soils. The end result is that sandy soils can withstand higher salinity irrigation water because more dissolved salts will be removed from the root zone by leaching.

Another important aspect of soil texture has to do with surface area. Because of their tiny size, a given volume of clay particles has far more surface area than the same volume of a larger sized particle. This simply means that clay soils are at a greater risk than course textured soils for excess sodium to bind to them and cause dispersion. Sands have larger particle sizes, resulting in less surface area; correspondingly, they cannot accept as much sodium as clay particles.

Role of Clay Type

The three main clay types are montmorillonite, illite, and kaolinite clays. On the microscopic scale, each of these clays has a different lattice structure, i.e., different building blocks. This directly affects the ability of sodium to bind to each type. Basically, the more sodium a certain type of clay is able to hold, the more infiltration and hydraulic conductivity will be reduced. Montmorillonite clays are affected by sodium the most, while kaolinite is least affected. This same pattern is also true for the swelling factor. Montmorillonites are the most prone to swelling and dispersion, whereas kaolinites are the least likely to swell and disperse.

Soil Salinity Measurements and Testing

Problems due to soil salinity and sodicity in soil are commonly evaluated by laboratory testing. The following laboratory measurements are typically used to determine the extent of these problems:

1. Electrical Conductivity (EC) – Measures the ability of the soil solution to conduct electricity and is expressed in decisiemens per meter (dS/m, which is equivalent to mmhos/cm). Because pure water is a poor conductor of electricity, increases in soluble salts result in proportional increases in the solution EC. The standard procedure for salinity testing is to measure EC of a solution extracted from a soil wetted to a "saturation paste." According to U.S. Salinity Laboratory Staff (1954), a saline soil has an EC of the saturated paste extract of more than 4 dS/m, a value that corresponds to approximately 40 mmol salts per liter. Crops vary in their tolerance to salinity and some may be adversely affected at ECs less than 4 dS/m. Salt tolerances are known for common crops. For example, peach is sensitive, whereas cotton is more salt tolerant (Maas, 1990). Beets and asparagus are very tolerant of salinity.

2. Total Soluble Salts (TSS) – Refers to the total amount of soluble salts in a soil-saturated paste extract expressed in parts per million or milligrams per liter (ppm or mg/L). A linear relationship exists between TSS and EC within a certain range that can be useful to closely estimate soluble

salts in a soil solution or extract. The ratio of TSS to EC of various salt solutions ranges from 550 to 700 ppm per dS/m. Sodium chloride, the most common salt, has a TSS of 640 ppm per dS/m. So if EC is known, TSS can be estimated using the formula below:

TSS (mg/L or ppm) = EC (mmhos/cm or dS/m) x 640

3. Sodium Adsorption Ratio (SAR) – A widely accepted index for characterizing soil sodicity, which describes the proportion of sodium to calcium and magnesium in soil solution. The formula to calculate SAR is given below, with concentrations expressed in milliequivalents per liter (meq/L) analyzed from a saturated paste soil extract.

$$SAR = \frac{\left[Na^+\right]}{\sqrt{1/2\left(\left[Ca^{2+}\right]+\left[Ma^{2+}\right]\right)}}$$

When SAR is greater than 13, the soil is called a sodic soil. Excess sodium in sodic soils causes soil particles to repel each other, preventing the formation of soil aggregates. This results in a very tight soil structure with poor water infiltration, poor aeration and surface crusting, which makes tillage difficult and restricts seedling emergence and root growth.

4. Exchangeable Sodium Percentage (ESP) – Another index that characterizes soil sodicity. Excess sodium causes poor water movement and poor aeration. By definition, sodic soil has an ESP greater than 15 (US Salinity Lab Staff, 1954). ESP is the sodium adsorbed on soil particles as a percentage of the Cation Exchange Capacity (CEC). It is calculated as:

$$ESP = \frac{\left[Na^+\right]}{CEC} \times 100$$

CEC is often estimated as the sum of the major exchangeable cations, including hydrogen. Both cations and CEC are expressed as meq/100g. ESP can also be calculated as:

$$ESP = \frac{\left[Na^+\right]}{\left[Ca^{2+} + Ma^{2+} + Na^+ + K^+\right]} \times 100$$

ESP is used to characterize the sodicity of soils only, whereas SAR is applicable to both soil and soil solution or irrigation water.

The Natural Resources Conservation Service (NRCS) provides the following classification of salt-affected soils using the saturated paste extraction:

Class	EC(mmhos/cm)	SAR	ESP	Typical soil structural condition*
Normal	Below 4.0	Below 13	Below 15	Flocculated
Saline	Above 4.0	Below 13	Below 15	Flocculated
Sodic	Below 4.0	Above 13	Above 15	Dispersed
Saline-Sodic	Above 4.0	Above 13	Above 15	Flocculated
*Soil structural condition also depends on other factors not included in the NRCS classification system, including soil organic matter, soil texture and EC of irrigation water.				

Collecting Soil Samples for Salinity Testing

The goal of salinity testing is to determine the salt level of soil from which roots extract water. Therefore, soil samples should be collected from the 0 to 6 inch depth or from the rooting depth. Deeper samples may be collected if the goal is to identify the extent of salinity caused by irrigation within the soil profile. In many cases, comparing soil samples from the affected area to surrounding normal-looking areas is valuable in diagnosing the problem. Collect eight to 10 cores from around a uniform area, mix them in a clean plastic bucket and transfer a composite sample (approximately 1 pound) to a soil sample bag.

Soil Salinity Test Package offered at UGA

Basic Soil Salinity Test (1:2 soil-to-water ratio): Because methods of obtaining soil solution samples at water contents more typical of the field condition (the saturated paste) are not very practical, aqueous extracts of the soil samples have traditionally been made in the laboratory at higher-than-normal water contents for routine soil salinity diagnosis and characterization. However, the values from the 1:2 soil-to-water extraction are different from the standard saturated paste extraction. Therefore, the interpretation shown below would also be different. The following parameters are included in the basic soil salinity test.

Calcium (Ca^{2+})

- Magnesium (Mg^{2+})

- Potassium (K^+)

- Sodium (Na^+)

- pH

- Electrical Conductivity (EC)

- Total Soluble Salts (TSS)

- Sodium Adsorption Ratio (SAR)

The University of Georgia adopted the following guidelines in interpreting EC data from soil extracts (1:2 soil-to-water ratio).

Electrical Conductivity (mmhos/cm)	Rating	Interpretation
0 - 0.15	Very low	Plants may be starved of nutrients.
0.15 - 0.50	Low	If soil lacks organic matter. Satisfactory if soil is high in organic matter.
0.51 - 1.25	Medium	Okay range for established plants.
1.26 - 1.75	High	Okay for most established plants. Too high for seedlings or cuttings.
1.76 - 2.00	Very high	Plants usually stunted or chlorotic.
>2.00	Excessively high	Plants severely dwarfed; seedlings and rooted cuttings frequently killed.

Comprehensive Soil Salinity Test

(Saturated Paste Extract; NH$_4$OAc Extraction for ESP)

The saturated paste method provides a more representative measurement of total soluble salts in the soil solution because it is closer to the water content of the soil under field conditions. This method, however, is time-consuming and subject to some technician variability when preparing the paste. This test package also includes extraction of soils with ammonium acetate (NH$_4$OAc) to measure exchangeable cations and ESP.

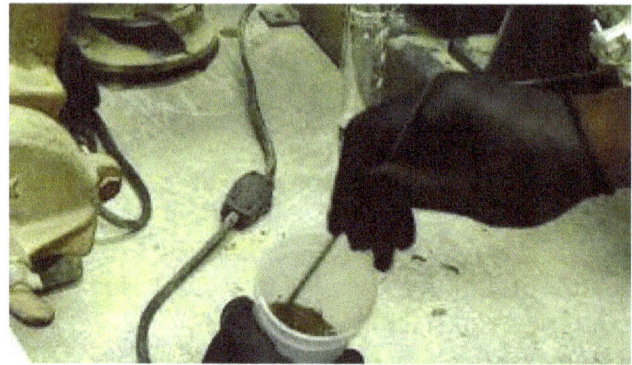

Preparing saturated pastes.

The following parameters are included in the comprehensive soil salinity test:

1. Saturated Paste Extract

- Calcium (Ca^{2+})

- Magnesium (Mg^{2+})

- Potassium (K$^+$)

- Chloride (Cl$^-$)

- Sodium (Na$^+$)

- Nitrate-nitrogen (NO$_3$-N)

- Sulfate (SO$_4^{2-}$)

- pH

- Electrical Conductivity (EC)

- Total Soluble Salts (TSS)

- Sodium Adsorption Ratio (SAR)

2. NH$_4$OAc Extract

- Calcium (Ca^{2+})

- Magnesium (Mg^{2+})

- Sodium (Na$^+$)

- Potassium (K$^+$)

- Exchangeable Sodium Percentage (ESP)

The general interpretation of data on electrical conductivity and exchangeable sodium percentage from saturated paste extracts are given below:

Interpretation of electrical conductivity (EC) from saturated paste extract.		
Electrical conductivity (mmhos/cm)	Salt Rank	Interpretation
0-2	Low	Very little chance of injury on all plants.
2-4	Moderate	Sensitive plants and seedlings of others may show injury.
4-8	High	Most non-salt tolerant plants will show injury; salt-sensitive plants will show severe injury.
8-16	Excessive	Salt-tolerant plants will grow; most others show severe injury.
16+	Very Excessive	Very few plants will tolerate and grow.

Interpretation of exchangeable sodium percentage (ESP) from saturated paste extract.		
ESP	Rank	Interpretation
0-10	Low	No adverse effect on soil is likely.
10+	Excessive	Soil dispersion resulting in poor soil physical condition and poor plant growth are likely.

Reclaiming Salt-affected Soils

After the kinds and amounts of salts in the soil have been determined by testing, the soil can be properly treated. Reclaiming a salt-affected soil involves leaching, chemical treatment or a combination of both.

Leaching. Application of good quality irrigation water in the correct amounts will remove excess salts from soils that are well structured and have good internal drainage. Excess salts should be leached below the root zone so that the EC of the soil solution becomes lower than the crop's critical threshold. The University of Georgia recommends leaching techniques to remove salts from the root zone when EC is greater than 1.25 mmhos/cm at a soil-to-water ratio of 1:2. The volume of low-salt water needed to dissolve and leach any large quantities of salts from the soil is given below. A general rule of thumb is that 6 inches of water will remove about half of the salt, 12 inches will remove four-fifths of the salt, and 24 inches will remove nine-tenths of the salt.

Estimated leaching requirements to remove salts	
Volume of salt-free water	Reduction of salt content in soil
6 inches	50%
12 inches	80%
24 inches	90%

For soils with poor drainage, it is recommended to break root-restrictive hardpans or clay pans by deep tillage to allow water to penetrate and leach the salts. It may be necessary to install tile drains to remove salt-laden drainage water and move it below the root zone by rainfall or irrigation water.

Chemical Treatment

When a soil has an SAR value of above 13 (or ESP greater than 15), it contains excess sodium that makes it a sodic soil. Excess sodium can cause soil dispersion, which prevents the formation of soil aggregates, resulting in surface sealing or crusting. Dispersion of the soil by excess sodium reduces water infiltration and movement through the soil, and also causes poor aeration. Good aeration and water movement are both essential to unrestricted growth of plant roots. To eliminate surface sealing, the soil should be treated with calcium to remove sodium. One of the most commonly used calcium sources for correcting sodium-contaminated soil is gypsum (calcium sulfate, $CaSO_4 \cdot 2H_2O$). Gypsum is incorporated into the soil, followed by application of salt-free irrigation water. The amount of calcium to apply depends on the quantity of sodium in the soil. Road de-icing salt or calcium chloride ($CaCl_2$) is also an option to provide calcium to soil, but it is more expensive than gypsum.

The table below provides the amount of gypsum needed to replace exchangeable sodium in the soil. The soil test sodium values are determined from an ammonium acetate extraction at pH 7.

Soil exchange-able sodium	Amount of Gypsum Neededa ($CaSO_4 \cdot 2H_2O$)			
lbs/acre	Tons/acre-foot[b]	Tons/acre-6 inchesc	lbs/1000 ft²-1 foot	lbs/1000 ft²-6 inches
460	1.7	0.9	80	40
920	3.4	1.7	160	80
1380	5.2	2.6	240	120
1840	6.9	3.4	320	160
2300	8.6	4.3	400	200
2760	10.3	5.2	480	240

Correcting a salt-affected soil involves identifying the kind and amount of salt, chemical treatment and leaching. When a salinity problem is identified, it is recommended that corrective steps be taken immediately. Prompt action will give a better chance of reclaiming the affected soil, be less expensive and pose less risk of plant damage.

Soil Liming

Liming is a traditional procedure in preparing soil for planting. It is the application of calcium- and magnesium-rich materials to soil in various forms, including marl, chalk, limestone, or hydrated lime. Lime used on farm land is also called agricultural lime. The primary reason to apply agricultural lime is to correct the high levels of acidity in the soil. Acid soils reduces plant growth by inhibiting the intake of major plant nutrients -nitrogen, phosphorus and potassium. Some plants, for example legumes, will not grow in highly acidic soils.

Soils become acidic in a number of ways:

- Leaching of land caused by high rainfall levels.

- Minerals loss over time caused by crop removal.

- Application of modern chemical fertilizers, which are the major contributors of acidified soil.

Soil with pH below 5.5 and below 70% of saturation requires liming. The best time is when plowing stubble, when there are no crops in the field. Effect of liming takes on average 6-7 years.

Agricultural lime have good effects on soil:

- Increases the pH of acidic soil.

- Provides a source of calcium and magnesium for plants.

- Permits improved water penetration for acidic soils.

- Improves the uptake of major plant nutrients (nitrogen, phosphorus, and potassium) of plants growing on acid soils.

Most of farming crops require neutral soil, with pH around 6-7, but there are also cultures that need expressly acidic or alkaline soil.

Types of Liming Materials

Agricultural Lime (Calcium Carbonate)

This is the most commonly used liming material on the North Coast. It consists of limestone crushed to a fine powder and is usually the cheapest material for correcting soil acidity. Good quality lime has 37 - 40% calcium.

Burnt Lime (Calcium Oxide)

Also known as quicklime, burnt lime is derived by heating limestone to drive off carbon dioxide. It is more concentrated and caustic than agricultural lime and unpleasant to handle, so is rarely used in agriculture.

Hydrated Lime (Calcium Hydroxide)

This is made by treating burnt lime with water, and is used mainly in mortar and concrete. It is more expensive than agricultural lime.

Dolomite

Dolomite is a naturally occurring rock containing calcium carbonate and magnesium carbonate. Good quality dolomite has a NV of 95-98, and contains 22% calcium and 12% magnesium. It is good for acid soils where supplies of calcium and magnesium are low, but if used constantly may cause a nutrient imbalance, because the mix is two parts calcium to one part magnesium (2:1), whereas the soil ratio should be around 5:1.

Gypsum (Calcium Sulfate)

Gypsum is not considered as a liming material, as it does not reduce soil acidity. It is used mainly to improve the structure of sodic clay soils, and these are not common in many areas of Tasmania.

Rates of Lime to Apply

As soil acidity increases (the lower the pH), more lime is needed to ameliorate acidity. You will have to add more lime to clay soils and peaty soils than you will to sandy soils to achieve the same result because different soil types react in different ways to the application of lime. The amount of lime to apply depends on three main factors; neutralising value, fineness of the lime and soil texture.

Neutralising Value (NV)

NV tells you the lime's capacity to neutralise soil acidity. Pure calcium carbonate has NV of 100, which is the standard. Ideally, NV should be over 95. The NV figure is marked on the lime bag or the invoice if you buy bulk lime.

Fineness

The finer the particles of lime, the faster they react with soil. Lime manufacturers have to specify the percentages of different-sized particles in their product.

Soil Texture (Amount of Sand, Silt and Clay)

It is easier to change pH on a sandy soil than on a clay soil. The estimated pH increases over the upper 10 cm of soil due to the addition of 1t/ha (1 kg/10 sq metres) of 100%NV product to different soil types are:

- Sand 0.5 - 0.7

- Loam 0.3 - 0.5

- Clay 0.2 - 0.3

- Red clay loam (basalt) 0.04 - 0.1

It is best to apply at least 2.5 t/ha to get a good response. The upper limit for one application is 7.5 t/ha.

Lime has to be physically in contact with moist acid soil in order to neutralise acidity. Lime dissolves slowly in the soil, therefore, incorporation in the top 10cm of soil (or deeper if possible) is best to increase the rate of reaction and leaching of lime to a greater depth. Incorporating lime will increase soil pH in the 0-10cm soil depth within 1-3 years.

References

- Soil-chemistry, Teaching-Organic-Farming: ucsc.edu, Retrieved 19-February-2019

- Soil -reaction-types-factors-and-influence-soil-science, soil-reaction: soilmanagementindia.com, Retrieved August 11, 2019

- Soil -acidity-definition-sources-and-problems, soil-acidity: soilmanagementindia.com, Retrieved August 2, 2019

- Soil -prop, background, cbm, energy: montana.edu, Retrieved March 17, 2019

- Importance-of-liming-for-high-soil-fertility, post: agrivi.com, Retrieved January 7, 2019

- Soil -ph-liming, soil-management, land-management-and-soils, agriculture: tas.gov.au March 13, 2019

Chapter 5

Soil Biology and Ecology

The study of microbial and faunal activity and ecology in soil is termed as soil biology. It is primarily concerned with soil respiration, soil biological crust, soil regulators and soil system. Soil ecology is a domain that studies the interactions between soil biology and abiotic and biotic facets of the soil environment. The topics elaborated in this chapter will help in gaining a better perspective about the key concepts of soil biology and soil ecology.

Soil biology is the study of soil biota and the interactions they have with each other and their environment. Soil biota includes four broad groups, based on size:

- Microflora (e.g. Bacteria, archaea, fungi and viruses <5 μm),

- Microfauna (e.g. Nematodes 10 μm-2 mm and protozoa 5 μm - 200 μm),

- Mesofauna (e.g. Small arthropods like mites and collembola 100 μm - 2 mm),

- Macrofauna (e.g. Earthworms and insects 2 mm - 2 m).

The soil biota comprises an enormous diversity with reports suggesting there could be greater than 15 000 different species per gram of soil. Much of this diversity, largely from the microflora group, is yet to be classified, however genomic technologies are helping us identify previously unknown soil organisms. Soil biology is represented by all three kingdoms of life: Bacteria, Archaea and Eukarya; in nearly all evolutionary branches. With such phylogenetic diversity it goes hand in hand that soil biology also posses an incredible range of functional diversity.

There are three levels of participation by soil biota in natural soil processes/functions:

1. Ecosystem engineers (e.g. earthworms, termites & ants),

2. Litter transformers (e.g. microarthropods),

3. Micro-food webs (e.g. microbes and microfaunal predators).

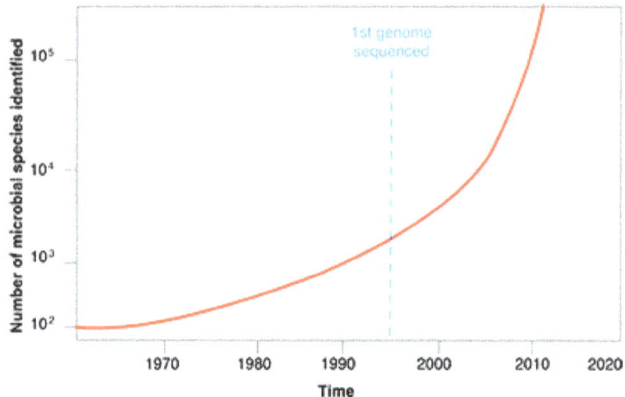

This diagram highlights how using genomic technologies based on DNA have exponentially increased the number of microbial organisms we have been able to identify.

Ecosystem Engineers

Ecosystem engineers such as ants and earthworms primarily alter the physical structure of soil but also have an influence on the overall rates of nutrient cycling and energy flows. These organisms initiate fragmentation of organic residues and take organic matter deeper into soil profiles. They also create pores that allow water and plant roots (as well as other soil biota) access to deeper parts of the profile.

Centipede

Litter Transformers

Litter transformers such as the microarthropod collembolas are involved in fragmenting plant residues and other organic substances making this material more available to microbes by increasing residue surface area for further chemical degradation and nutrient cycling.

Micro-food Web Processors

Soil biota play a major role at all levels in micro-food webs whereby energy is transferred from one organism to another and as a consequence nutrients are recycled through the soil environment. For example, bacteria, archaea and fungi decompose plant litter through enzyme degradation to sequester nutrients and these organisms are in turn eaten by predatory protozoa, nematodes and arthropods. This predation helps maintain specific populations and keeps the predator prey balance in check. At the same time, however, litter transformers are breaking down dead organic matter, including remains of biota from all levels of the food web.

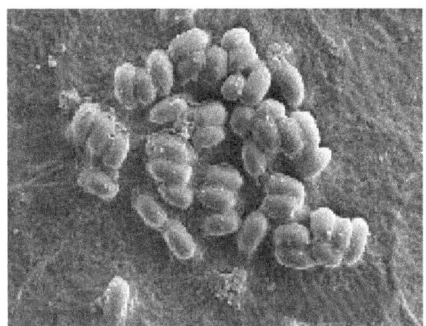

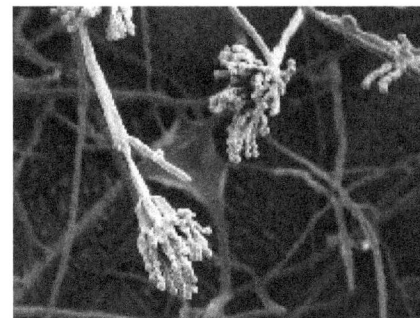

Bacterial colony   Fungal spores forming at the terminus of hyphae

Soil biology is not evenly distributed in soil and occurs in 'hot-spots' associated with soil organic matter. Decomposing residues, (obvious sites for both physical and enzymatic attack by a range of soil biota), the rhizosphere, (the soil zone that surrounds and is influenced by the roots of plants) and macroaggregates are all examples of such hot spots.

Soil Organisms

Soil organisms are responsible, to a varying degree depending on the system, for performing vital functions in the soil. Soil organisms make up the diversity of life in the soil. This soil biodiversity is an important but poorly understood component of terrestrial ecosystems. Soil biodiversity is comprised of the organisms that spend all or a portion of their life cycles within the soil or on its immediate surface (including surface litter and decaying logs).

Soil organisms represent a large fraction of global terrestrial biodiversity. They carry out a range of processes important for soil health and fertility in soils of both natural ecosystems and agricultural systems. This annex provides brief descriptions of organisms that are commonly found in the soil and their main biological and ecological attributes.

The community of organisms living all or part of their lives in the soil constitute the soil food web. The activities of soil organisms interact in a complex food web with some subsisting on living plants and animals (herbivores and predators), others on dead plant debris (detritivores), on fungi or on bacteria, and others living off but not consuming their hosts (parasites). Plants, mosses and some algae are autotrophs, they play the role of primary producers by using solar energy, water and carbon (C) from atmospheric carbon dioxide (CO_2) to make organic compounds and living tissues. Other autotrophs obtain energy from the breakdown of soil minerals, through the oxidation of nitrogen (N), sulphur (S), iron (Fe) and C from carbonate minerals. Soil fauna and most fungi, bacteria and actinomycetes are heterotrophs, they rely on organic materials either directly (primary consumers) or through intermediaries (secondary or tertiary consumers) for C and energy needs.

A food-web diagram shows a series of conversions (represented by arrows) of energy and nutrients as one organism eats another. The "structure" of a food web is the composition and relative numbers of organisms in each group within the soil. The living component of soil, the food web, is complex and has different compositions in different ecosystems. In a healthy soil, there are a large number of bacteria and bacterialfeeding organisms. Where the soil has received heavy treatments of pesticides, chemical fertilizers, soil fungicides or fumigants that kill these organisms, the beneficial soil organisms may die (impeding the performance of their activities), or the balance between the pathogens and beneficial organisms may be upset, allowing those called opportunists (disease-causing organisms) to become problems.

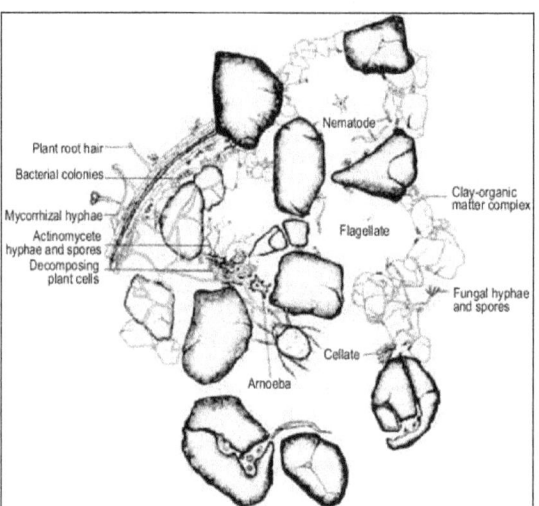

The soil environment

Table: Categories and characteristics of soil organisms

Category	Characteristics	Organisms
Permanent	Whole life cycle in the soil	Mites, collembola, earthworms
Temporal	Part of life cycle in the soil	Insect larvae
Periodical	Frequently enter into the soil	Some insect larvae
Transitory	An inactive phase in the soil (e.g. eggs, pupae, hibernation) but the active period not in the soil	Some insects
Accidental	Organisms fall down or they are drawn along	Insect larvae

The easiest and most widely used system for classifying soil organisms is by using body size and dividing them into three main groups: macrobiota, mesobiota and microbiota. The ranges that determine each size group are not exact for all members of each group.

Micro-organisms

These are the smallest organisms (<0.1 mm in diameter) and are extremely abundant and diverse. They include algae, bacteria, cyanobacteria, fungi, yeasts, myxomycetes and actinomycetes that are able to decompose almost any existing natural material. Micro-organisms transform organic matter into plant nutrients that are assimilated by plants. Two main groups are normally found in agricultural soils: bacteria and mycorrhizal fungi.

Bacteria

Bacteria are very small, one-celled organisms that can only be seen with a powerful light (1 000×) or electron microscope. They constitute the highest biomass of soil organisms. They are adjacent and more abundant near roots, one of their food resources. There are many types of bacteria but the focus here is on those that are important for agriculture, e.g. Rhizobium and actinomycetes.

Bacteria are important in agricultural soils because they contribute to the carbon cycle by fixation (photosynthesis) and decomposition. Some bacteria are important decomposers and others such as actinomycetes are particularly effective at breaking down tough substances such as cellulose (which makes up the cell walls of plants) and chitin (which makes up the cell walls of fungi). Land management has an influence on the structure of bacterial communities as it affects nutrient levels and hence can shift the dominance of decomposers from bacterial to fungal.

One group of bacteria is particularly important in nitrogen cycling. Free-living bacteria fix atmospheric N, adding it to the soil nitrogen pool; this is called biological nitrogen fixation and it is a natural process highly beneficial in agriculture. Other Nfixing bacteria form associations (in the form of nodules) with the roots of leguminous plants.

The nodule is the place where the atmospheric N is fixed by bacteria and converted into ammonium that can be readily assimilated by the plant. The process is rather complicated but, in general, the bacteria multiply near the root and then adhere to it. Next, small hairs on the root surface curl around the bacteria and they enter the root. Alternatively, the bacteria may enter directly through points on the root surface. Once inside the root, the bacteria multiply within thin threads. Signals stimulate cell multiplication of both the plant cells and the bacteria. This repeated division results in a mass of root cells containing many bacterial cells. Some of these bacteria then change into a form that is able to convert gaseous N into ammonium nitrogen (they can "fix" N). These bacteria are then called bacteroids and present different properties from those of free cells. Most plants need very specific kinds of rhizobia to form nodules. A specific Rhizobium species will form a nodule on a specific plant root, and not on others. The shapes that the nodules form are controlled by the plant and nodules can vary considerably in size and shape.

- Rhizobium and the Nodulation Process

The nodulation process is a series of events in which rhizobia interact with the roots of legume plants to form a specialized structure called a root nodule.

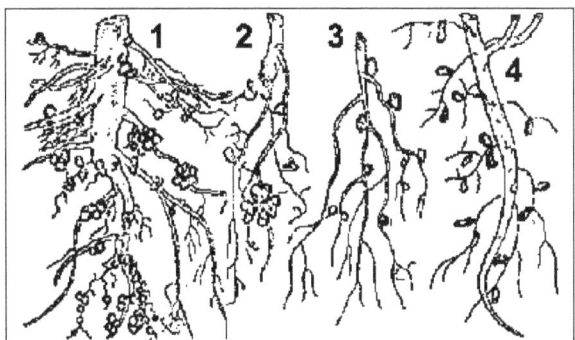

Different types of nodules on leguminous roots:
(1) soybean; (2) alfalfa; (3) pea; and (4) white clover

Actinomycetes are a broad group of bacteria that form thread-like filaments in the soil. The distinctive scent of freshly exposed, moist soil is attributed to these organisms, especially to the nutrients they release as a result of their metabolic processes. Actinomycetes form associations with some non-leguminous plants and fix N, which is then available to both the host and other plants in the near vicinity.

Bacteria produce (exude) a sticky substance in the form of polysaccharides (a type of sugar) that helps bind soil particles into small aggregates, conferring structural stability to soils. Thus, bacteria are important as they help improve soil aggregate stability, water infiltration, and water holding capacity. However, in general their effect is less marked than that originated by large invertebrates such as earthworms.

Fungi

These organisms are responsible for the important process of decomposition in terrestrial ecosystems as they degrade and assimilate cellulose, the component of plant cell walls. Fungi are constituted by microscopic cells that usually grow as long threads or strands called hyphae of only a few micrometres in diameter but with the ability to span a length from a few cells to many metres. Soil fungi can be grouped into three general functional groups based on how they source their energy:

- Decomposers - saprophytic fungi - convert dead organic material into fungal biomass, CO_2, and small molecules, such as organic acids. These fungi generally use complex substrates, such as the cellulose and lignin, in wood. They are essential for decomposing the carbon ring structures in some pollutants. Like bacteria, fungi are important for immobilizing or retaining nutrients in the soil.

- Mutualists - mycorrhizal fungi - colonize plant roots through a symbiotic relationship. The definition of symbiosis is a close, prolonged association between two or more different organisms of different species that may benefit each member. Mycorrhizae increase the surface area associated with the plant root, which allows the plant to reach nutrients and water that otherwise might not be available. Mycorrhizae essentially extend plant reach to water and nutrients, allowing plants to utilize more of the resources available in the soil. Mycorrhizae source their carbohydrates (energy) from the plant root they are living in/on and they usually help the plants by transferring phosphorus (P) from the soil into the root. Two major groups are identified: (i) ectomycorrhizae, that grow on the surface layers of the roots and are commonly associated with trees; and (ii) endomycorrhizae, such as arbuscular mycorrhizal fungi and vesicular mycorrhizal fungi, that grow within the root cells and are commonly associated with grasses, row crops, vegetables and shrubs. Arbuscular mycorrhizal fungi can also benefit the physical characteristics of the soil because their hyphae form a mesh to help stabilize soil aggregates. Vesicular-arbuscular mycorrhizae are the most widespread mycorrhizal fungi. Mycorrhizae are particularly important for phosphate uptake because P does not move towards plant roots easily. These organisms do not harm the plant, and in return, the plant provides energy to the fungus in the form of sugars. The fungus is actually a network of filaments that grows in and around the plant root cells, forming a mass that extends considerably beyond the root system of the plant.

Pathogens or parasites cause reduced production or death when they colonize roots and other organisms. Root-pathogenic fungi, such as Verticillium, Pythium and Rhizoctonia, cause major economic losses in agriculture each year. Many fungi help control diseases, e.g. nematode-trapping fungi that parasitize disease-causing nematodes, and fungi that feed on insects may be useful as biocontrol agents.

Microfauna

The microfauna (<0.1 mm in diameter) includes inter alia small collembola and mites, nematodes and protozoa that generally live in the soil water films and feed on microflora, plant roots, other microfauna and sometimes larger organisms (e.g. entomopathogenic nematodes feed on insects and other larger invertebrates). They are important to release nutrients immobilized by soil microorganisms.

Nematodes

Nematodes are tiny filiform roundworms that are common in soils everywhere. They may be free-living in soil water films; beneficial for agriculture or phytoparasitic, and live at the surface or within the living roots (parasites). Free-living nematodes graze on bacteria and fungi, thus they control the populations of harmful micro-organisms. These nematodes are 0.15-5 mm long and 2-100 ìm wide; an exception are Mermithidae nematodes, which may be 20 cm long and are very common in tropical soils, being parasites of some arthropods such as locusts. Nematodes can only move through the soil where a film of moisture surrounds the soil particles. They live in the water (they are hydrobionts) that fills spaces between soil particles and covers roots. In hot and dry conditions, they enter into a dormant stage, and as soon as water becomes available, they spring back to activity.

Nematodes are recognized as a major consumer group in soils, generally grouped into four to five trophic categories based on the nature of their food, the structure of the stoma (mouth) and oesophagus, and the method of feeding bacterial feeders, fungal feeders, predatory feeders, omnivores, and plant feeders. The bacterial feeders prey on bacteria (bacterivores) and may ingest up to 5 000 cells/minute, or 6.5 times their own weight daily. This helps disperse both the organic matter and the decomposers in the soil. Bacterial- and fungal-feeding nematodes release a large percent of N when feeding on their prey groups and are thus responsible for much of the plant available N in the majority of soils. The annual overall consumption may be as much as 800 kg of bacteria per hectare and the amount of N turned over in the range of 20-130 kg.

Phytophages or plant-feeding nematodes damage plant roots, with important economic consequences for farmers. They possess stylets with a wide diversity of size and structure, and they are the most extensively studied group of soil nematodes because of their ability to cause plant disease and reduce crop yield.

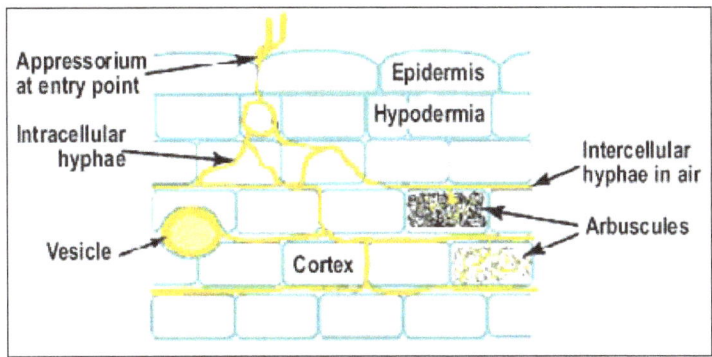

Mycorrhizal roots and the associated networks
of hyphae are a major component of most soils

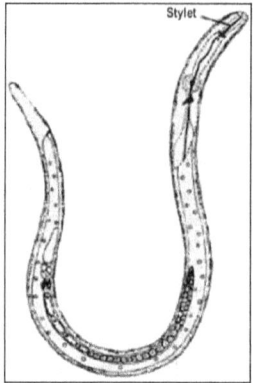

Drawing of a plant-parasitic nematode
showing the stylet used to penetrate plant roots

Mesofauna

Mesofauna (0.1-2 mm in diameter) includes mainly microarthropods, such as pseudoscorpions, springtails, mites, and the worm-like enchytraeids. Mesofauna have limited burrowing ability and generally live within soil pores, feeding on organic materials, microflora, microfauna and other invertebrates.

A drawing of a "springtail"

Collembola

Collembola or "springtails" are microarthropods that live in the litter or in the pore space of the upper 10-15 cm of soil. They are saprophagous and feed mainly on fungi, bacteria and algae growing on decomposing plant litter. They are important as epigeic decomposers. Unlike most insects, they have no wings at any stage. They measure a few millimetres in length and elongate with a characteristic salutatory organ, a forked "tail" which enables them to spring when in danger. Springtails are probably the most abundant group of insects on Earth.

Pseudoscorpions

Pseudoscorpions are tiny arachnids rarely longer than 8 mm. They live in litter, decaying vegetation, and the soil. Pseudoscorpions superficially resemble true scorpions, bearing relatively large chelae on the pedipalps, but they do not have a telson or stinger. Pseudoscorpions feed on very small arthropods such as springtails and mites.

- Macrofauna

Members of species classed as macrofauna are visible to the naked eye (generally> 2 mm in diameter). Macrofauna includes vertebrates (snakes, lizards, mice, rabbits, moles, etc.) that primarily dig within the soil for food or shelter, and invertebrates (snails, earthworms and soil arthropods such as ants, termites, millipedes, centipedes, caterpillars, beetle larvae and adults, fly and wasp larvae, spiders, scorpions, crickets and cockroaches) that live in and feed in or upon the soil, the surface litter and their components. In both natural and agricultural systems, soil macrofauna are important regulators of decomposition, nutrient cycling, soil organic matter dynamics, and pathways of water movement as a consequence of their feeding and burrowing activities. Here the focus is on soil invertebrates.

- Earthworms

The effects of earthworms in the soil differ according to the ecological category of the species involved:

Epigeic: they live in the litter layers, a very changing environment, subject to drought, high temperatures and predator presence. These earthworms are generally small and pigmented (green or reddish) with rapid movements.

Endogeic: these are unpigmented (with no colour) worms that live and feed in the soil. This group is further divided into three subgroups: oligohumic, mesohumic and polyhumic, depending on the organic matter content of the soil ingested.

Anecic: these earthworms feed on the surface litter that they generally mix with soil, but they spend most of their time in the soil. They are large, with dark anterodorsal pigmentation, and they dig subvertical burrows.

As a result of this wide range of adaptations, earthworms have diverse functions in the soil. Epigeic worms can be used as compost makers with no impact on soil structure. Anecics and endogeics do have an impact on soil structure owing to their mixing and burrowing activities, and on soil organic matter.

Earthworms generally exert beneficial effects on plant growth. However, negative effects may be induced under particular situations. The effect on grain yields is also proportional to the earthworm biomass; significant effects start to appear at biomass values> 30 g fresh weight, although very high biomasses of single species of earthworms, (e.g. Pontoscolex corethrurus) may inhibit production under particular situations. Many mechanisms are involved in the growth stimulation. These vary from large-scale effects on soil structure and nutrient availability, to the enhancement of mycorrhizal infection or control of plant-parasitic nematodes. Once the earthworms are established, a dynamic cropping system - involving crop rotations with long cycle crops or perennials with good organic matter additions - contributes to securing long-lasting benefits from earthworm activities.

Termites

Termites are important members of soil macrofauna in various regions of the world. They are social insects, living in organized colonies with a number of castes (different individuals) with a set of morphological and physiological specializations. The main castes are: queen (the termite that founds the colony), worker and soldier.

Neither individual termites nor colonies normally travel long distances as they are constrained to live within their territorial border or within their food materials. A number of species feed on living plants and some may become serious pests in agricultural systems where dead residues are scarce. Most species feed on dead-plant materials above and below the soil surface. Their food sources include plant-decaying materials, dead foliage, woody materials, roots, seeds and the faeces of higher animals. There are also soil-wood feeders and soil feeders, which means that they ingest a high proportion of mineral material. Their nutrition derives mainly from well-decayed wood and partly humified soil organic matter. Another group of termites grow fungi in their nests (fungus-growing termites).

Termites may be classified by their feeding habits:

- Grass harvesters,

- Surface litter feeders,

- Wood feeders,

- Soil-wood feeders,

- Soil feeders (humivores).

The nests formed by termites may occur in different locations, e.g. within the wood of living or dead trees, in subterranean locations, in other nests formed by other termite species, and by forming epigeal nests (above the soil surface) and arboreal nests.

Ants

Ants build a large variety of structures in the soil. However, because of their feeding habits, they are of less importance in regulating processes in the soil than termites or earthworms.

Beetles

Beetles (Coleoptera) are diverse taxonomically and differ widely in size and in the ecological role they perform in soil and litter. They are either saprophagous, phytophagous or predators. Two groups are of particular relevance in agricultural soils: larvae from the family Scarabeidae (dung-beetles), crucial to burying cowdung in natural savannahs and grasslands used for cattle grazing (e.g. in Africa); and Melolonthinae beetles, whose larvae may be abundant in grasslands and affect crop production by feeding on living roots.

Dung beetles dig subvertical galleries 10-15 mm wide down to a depth of 50-70 cm with a variable number of chambers, which are further filled with large pellets of dung. The adult beetle lays one egg in each chamber and then the larva feeds on the pellet to complete the cycle (Lavelle and Spain, 2001). They generally give rise to small mounds a few centimetres high on the soil surface (Hurpin, 1962).

Biogenic Structures

Biogenic structures are those structures created biologically by a living organism. Three main groups of biogenic structures are commonly found in agricultural systems: earthworm casts and burrows, termite mounds and ant heaps. The biogenic structures can be deposited in the soil surface and in the soil, and generally they have different physical and chemical properties from the surrounding soil. The colour, size, shape and general aspect of the structures produced by large soil organisms can be described for each species that produces it. The form of the biogenic structure can be likened to simple geometric forms in order to facilitate evaluation of the volume of soil moved through each type of structure on the soil surface.

Earthworm Casts

Earthworm casts vary in size depending on the size of the earthworm that produces them. They range from a few millimetres to several centimetres in diameter, weighing from only a few grams to more than 400 g.

Granular casts are very small and formed by isolated faecal pellets. These casts can be found on the soil surface or within the soil, and are generally produced by epigeic earthworms.

Globular casts are larger and formed by large aggregates. These are normally produced by endogeic and anecic earthworms. The casts produced by anecic earthworms comprise an accumulation of somewhat isolated round or oval-shaped pellets (one to several millimetres in diameter) which may coalesce into "paste-like slurries" that form large structures. Hence, casts are large in size, towerlike, and made of superposed layers of different ages, the older (i.e. dry and hard) located at the base and the younger (i.e. fresh and soft) on the top. Casts produced by anecic earthworms have a higher proportion of organic matter, especially large particles of plant material, and a larger proportion of small mineral components than in the surrounding soil.

Plate.Granular casts on the soil surface of an African soil.

Globular casts deposited by an African earthworm.

Earthworm Burrows

Earthworms construct burrows or galleries through their movement in the soil matrix. The type and size of the galleries depends on the ecological category of earthworm that is producing it.

Anecic earthworms create semi-permanent subvertical galleries, while endogeic worms dig rather horizontal burrows. These galleries may be filled with casts, which can be split into smaller aggregates by other smaller earthworms or soil organisms. Galleries are cylindrical and their wall area coated with cutaneous mucus each time the worm passes through.

Soil micro-organisms (bacteria) are markedly concentrated at the surface of the gallery walls and within the adjacent 2 mm of the surrounding soil. This microenvironment comprises less that 3 percent of the total soil volume but contains 5-25 percent of the whole soil microflora and is where some functional groups of bacteria predominate.

An earthworm gallery (Martiodrilus sp.) in the Colombian savannah filled with casts and a root following the pathway opened by the earthworm. Root hairs are attached to the cast where higher availability of nutrients (C, N and P) exists compared with the surrounding soil.

Termite Mounds

Termite mounds are among the most conspicuous structures in savannah landscapes. Termite mounds are of diverse types and are the epigeal part of a termite nest that originates from subterranean beginnings. In Africa, termites build up half of the biomass of the plains.

Termites process high quantities of material in their building activities, thus influencing the soil properties as compared with surrounding soils. Soil texture and structure are modified strongly in termite mounds. In general, the soil of the termite mounds exhibits a higher proportion of fine particles (clay), which termites transport from the deeper to upper soil horizons.

Ant Heaps

The tropical American genera Acromyrmex and Atta leaf-cutting ants (Plate A1.4) make subterranean nests and their leaf harvesting may lead to enormous incorporations of organic matter and, hence, nutrients into the soil.

Roots

Although not generally considered soil organisms, roots grow mostly within the soil and have wide-ranging, long-lasting effects on both plant and animal populations aboveground and belowground, and thus they are included among soil biota.

Rhizosphere

The rhizosphere is the region of soil immediately adjacent to and affected by plant roots. It is a very dynamic environment where plants, soil, micro-organisms, nutrients and water meet and interact. The rhizosphere differs from the bulk soil because of the activities of plant roots and their effect on soil organisms.

The root exudates can be used to increase the availability of nutrients and they provide a food source for micro-organisms. This causes the number of microorganisms to be greater in the rhizosphere than in the bulk soil. Their presence attracts larger soil organisms that feed on micro-organisms and the concentration of organisms in the rhizosphere can be up to 500 times higher than in the bulk soil.

An important feature of the rhizosphere is the uptake of water and nutrients by plants. Plants take up water and nutrients into their roots. This draws water from the surrounding soil towards the roots and rhizosphere.

The soil organisms near the rhizosphere influence plant roots because:

- They alter the movement of C compounds from roots to shoots (translocation).

- Earthworm galleries (burrows) provide an easy pathway for roots to take as they grow through the soil (Plate A1.3).

- Micorrhizal associations can increase nutrient uptake by plants.

- Some of them are pathogenic and can attack plant roots.

Beneficial vs. Harmful Organisms in Agricultural Soils

Agricultural practices can have either positive or negative impacts on soil organisms. Land management and agricultural practices alter the composition of soil biota communities at all levels, with important consequences in terms of soil fertility and plant productivity. There are examples of both positive and negative effects of some groups of soil organisms, particularly micro-organisms, phytoparasites/pathogens or rhyzophages, plant roots, and macrofauna on plant production.

The different agricultural practices used by farmers also exert an important influence on soil biota, their activities and diversity. Clearing forested or grassland for cultivation has a drastic effect on the soil environment and, hence, on the numbers and kinds of soil organisms. In general, such activity reduces the quantity and quality of plant residues and the number of plant species considerably. Thus, the range of habitats and foods for soil organisms is reduced significantly. Through changing the physical and chemical environment, agricultural practices alter the ratio of different organisms and their interactions significantly, for example, through adding lime, fertilizers and manures, or through tillage practices and pesticide use.

Deposits of soil by a fungus-growing ant in the Colombian Llanos

The beneficial effects of soil organisms on agricultural productivity that may be affected include:

- organic matter decomposition and soil aggregation;

- breakdown of toxic compounds, both metabolic by-products of organisms and agrochemicals;

- inorganic transformations that make available nitrates, sulphates and phosphates as well as essential elements such as Fe and Mn;

- N fixation into forms usable by higher plants.

However, other soil organisms are detrimental or harmful to plant production. For example ants, aphids and phytophagous nematodes can be serious pests, and some micro-organisms, bacteria and actinomycetes cause also plant diseases. However, most damage is caused by fungi, which account for most soil-borne crop diseases.

Soil Regulators

There is growing awareness of the importance of soil biology to efficient soil function. It is therefore helpful to understand what regulates soil biological populations so that favourable management practices can be used to promote biological function.

There are two levels of regulation of soil biology. The primary regulators are environmental and include air, water, temperature, and soil type. In the context of a habitat all of these features are related such that the texture and porosity of a soil determines the water and air available for growth. The secondary regulators relate to organic matter quality and quantity, the amount and frequency of soil disturbance, and the inputs used to manage production (fertilisers, herbicides, lime etc.).

Primary Regulators

Soil Air

The movement of air into and out of the soil is critical to the survival of aerobic organisms, and the functions they perform. There exists therefore, a strong relationship between soil structure and soil biological functions.

Soil Water

Aerobic microbial activity responds to soil water potential assuming soil temperature is not limiting. Figure shows that microbial biomass is closely aligned with soil moisture content with populations rising with available moisture when food and temperature are not limiting. Above field capacity, the loss of soil oxygen due to inundation would see a reduction in microbial biomass.

Of the microorganisms, fungi are generally more tolerant of lower soil moisture than bacteria as bacteria are less mobile and rely on diffusion to obtain nutrients (NRCS, 2013b). Earthworms are generally numerous on dairy farms but populations fall off below 500-600mm annual rainfall with relatively few individuals remaining below 300mm.

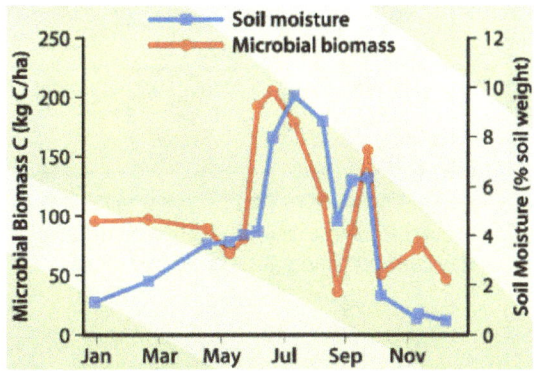

Microbial biomass activity as influenced by the level of soil moisture

Soil Temperature

Plants as well as soil organisms require certain minimum temperatures in order to grow and carry out their activities. Biological activity and associated growth and development occurs more quickly at higher temperatures, as evidenced by growth rates of pastures speeding up in spring and slowing as winter approaches.

In the soil, biological functions, such as breaking down of organic matter or cycling of nutrients, are similarly affected by temperature. This is why it is generally not a good idea to soil test for nitrogen in spring because the increase in biological activity releases nitrogen from stores of organic matter resulting in an inflated account of the true quantity of N in dairy soils.

Soil Type

Soil type strongly influences microbial populations for a number of reasons. Clay soils have the potential to hold more water and for longer than a sandy soil. For this reason, they generally hold more organic carbon than sandy soils. Well-structured clay soils will have a higher number of micro and macro-aggregates, thereby providing more potential habitat for soil organisms of varying sizes. A diversity of habitat will ensure maximum protection for soil organisms against predation.

Note that while clay soils have potential to support higher microbial biomass, this potential will only be realised if other regulators (primary and secondary) are not limiting. Most obviously this means that if the clay soil is poorly structured, its carbon capture potential, air-filled porosity, drainage, and numbers of micro-aggregates will be sub-optimal and production in such a soil is also likely to be below potential. So, even though clay soils have potential to support high microbial biomass, a heavy clay soil that has been used for cropping, or that was poorly managed in wet conditions may not be well structured and may have lower microbial biomass than a lighter loamy soil. Figure shows two different soil types under two different land uses. Higher clay, and less disturbance results in higher microbial biomass. Actual microbial biomass on individual farms will be strongly influenced by management practices.

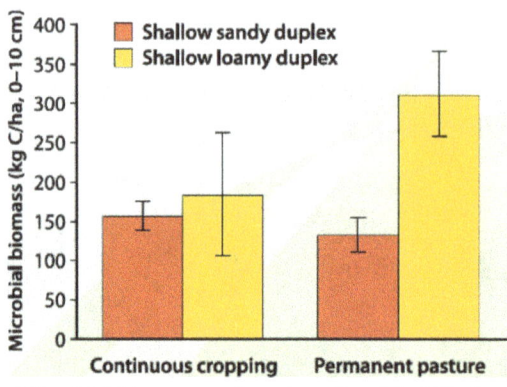

Microbial biomass carbon in soils with differing
clay contents and management practices

Survival Strategies

Bacteria are simple organisms consisting of a single prokaryotic (no cell nucleus) cell. They are extremely responsive to changes in their environment either rapidly dying back or reproducing at a very high rate depending on conditions. Under favourable conditions bacteria may divide every 20 minutes. This could result in exponential growth where one bacterium could produce 10 million in just 10 hours. However, this is unlikely to occur to this degree in soils due to reduction in food supply or accumulation of metabolic wastes. When conditions change to be less favourable, most bacteria can quickly develop a range of 'resting bodies' which can survive extended periods until such time as conditions again favour growth and development.

Fungi usually have plant-like vegetative bodies called mycelia (singular mycelium). The mycelium consists of elongated, branched, microscopic filaments termed hyphae. They are higher order eukaryotic (possessing a membrane-bound nucleus) organisms, the vast majority of which are saprophytic, that is they live on dead organic matter. Fungi reproduce primarily by means of spores. Fungi are not as responsive to environmental changes as bacteria due to their larger physical size allowing access to a wider range of soil resources. However spores may be produced as resting bodies when unfavourable conditions persist. Fungi may regrow from severed hyphae resulting from tillage but their recovery is slower than that of bacteria – food, water, air and nutrition notwithstanding.

Secondary Regulators

Organic Matter Quality and Quantity

Organic amendments: The carbon to nitrogen ratio (C:N) is a good measure of likely mineralisation (release of N by microbes) or immobilisation (tie-up of N in microbial biomass) in soil. A soil's C:N should be in the order of 12:1. If the C:N is greater than 25:1, immobilisation of nutrients is likely. If the C:N is less than 25:1, mineralisation is likely. This means that if microbes (mainly bacteria and archaea) are in a high nitrogen environment (low C:N) they can use that nitrogen to breakdown organic matter in soils to access nutrients, including carbon, as a food resource, and their population will likely increase thereby turning over N for plant access. However, in low C:N soils, soil carbon is at risk of declining if sufficient carbon is not re-introduced into the system by growth (e.g. plant roots), or application (e.g. manures or compost). In low C:N soils, microbes will use available nitrogen to degrade soil carbon which is released as CO_2

Conversely, in a high C:N soil, bacteria will access all available nitrogen to breakdown excess C and as they are superior competitors for soil N compared to plant roots (Owen & Jones, 2001), the plant will be deprived of nitrogen because it is immobilised in the bodies of bacteria and other soil microorganisms. This is termed 'nitrogen draw down'. It is usually a temporary phenomenon and is overcome when bacteria die off due to resource depletion, or another nitrogen source is introduced.

The C:N concept is important when adding organic amendments to soil. Table shows average C:N ratios of common organic materials. As mentioned above, if an amendment has a C:N of less than 25:1 it will progressively release nutrients and should have a fertiliser effect. If the material has a C:N ratio of more than 25:1, it is likely that nitrogen will be immobilised and nitrogen draw down will occur.

C:N ratios of common organic amendments:

MATERIAL	C:N
Urine	2:1
Dried blood	4:1
Pig manure	5:1
Poultry manure	10:1
Farmyard manure	14:1
Seaweed	19:1
Horse manure	25:1
Weeds	30:1
Straw	80:1
Sawdust	500:1

Microbial diversity: Microbial diversity refers to the number and variety of soil microorganisms. As a general rule, the more diverse the above-ground crop or pasture mix, the more diverse will be the micro-biological communities in the soil. Wilhelm cites the 'Elton Principle' which holds that the greater the complexity of a microbiological community in terms of total number and species of organisms, the greater the stability of the community. As shown in table, organic amendments can vary considerably and their application to the soil will have different effects on the soil biological community. In the same way, crop and pasture mixes will also influence microbial composition and activity.

Plants vary in the size and structure of their root systems, in the quantity and quality of root exudates, and in the degradability of crop or pasture residues. This results in differences in microbial density and diversity in the plant rhizosphere (root zone) and near crop and pasture residues. Some plants possess chemicals that inhibit the growth of other plants, or have negative effects on soil organisms. Likewise, some microbes possess strategies that enhance their competitive advantage. This has particular relevance when we consider suppressive soils. The term 'suppressive soil' has been used to describe soils in which a pathogen is present but is not causing economic damage. Suppression of a pest or disease is the mechanism by which one or several organisms are antagonistic to a pathogen through the antibiotics they produce, competition for food, or through direct parasitising of the pathogen. As an example, the production of isothiocyanates in canola has a suppressing effect on soil microorganisms. Isothiocyanates possess fungicidal, bacteriocidal, nematocidal and allelopathic properties.

Tillage

Organic matter persists in soils to the degree that it is protected from microbial attack or because prevailing moisture and temperature conditions are unfavourable for microbial decomposition. Tillage of any kind impacts on these protective mechanisms and renders the organic matter vulnerable to degradation by soil organisms. Tillage mixes the soil bringing microbes into more intimate contact with organic matter. It also improves (however temporarily) air and water movement into the soil – elements important for the growth and development of soil biology. Tillage can also be used to incorporate and distribute plant residue into the soil profile, again bringing food resources into close contact with soil organisms.

Tillage favours bacteria in view of their superior ability to respond quickly to changes in the environment. Fungal populations tend to be negatively impacted in view of the damage to the hyphal networks.

The incorporation of large quantities of organic material into soil can be a positive undertaking provided follow up actions maximise the use and sequestration potential of incorporated organic matter. For example, discing in crop residues will help to capture much of the carbon turned into the soil and will benefit the establishment and growth of perennial pasture.

The development of minimum- or no-till systems recognises the value of minimising soil disturbance. Stubble retention or surface applied organic materials will support slower decomposition and nutrient mineralisation, favour fungal growth to aerobically degrade lignocellulose compounds, promote better aggregation and soil structure, and improve the potential for SOM accumulation.

Chemical Impacts on Soil Biology

The large number of chemicals registered for use on farms makes it difficult to discriminate between those that are benign and those that are harmful to soil biology. While some have little effect, others do negatively impact on soil biology. Bunemann et al. reviewed the impact of agricultural impacts on soil organisms and found that:

- Fertilisers generally enhanced soil biological activity due to increases in production;

- The acidifying effects that can occur with the use of certain nitrogenous fertilisers resulted in negative impacts on soil biological activity;

- Organic amendments generally enhanced soil biological activity;

- Microbial inoculation, with the exception of nitrogen fixing microbes, appears to have little long-term effect;

- The negative effects of pesticides and fungicides were more commonly reported;

- Negative effects of herbicides were less commonly reported.

The negative impact on soil biological activity of many herbicides is reversible i.e. given sufficient time, the soil biology bounces back. However, with some chemicals, repeated applications delays or removes that reversibility. It recommended that:

- The short-term impacts of most of the herbicides tested are reversible, so it may be possible to develop management options to reduce non-target negative impacts;

- An appropriate recovery period for soil biota should be allowed between herbicide applications;

- Soils with a healthy biota could recover from short-term negative effects of herbicide application. Appropriate use of herbicides could be less destructive to soil biota if management practices that improve biological activity are promoted.

Grains Research and Development Council (GRDC) funded research from South Australia found specific effects of herbicides on soil N fixing bacteria with reductions in nodulation that resulted in reduced N benefit to the system.

Lime application and the associated increase in soil pH are strongly correlated with changes in microbial communities. Lime has also been shown to influence functioning of the nitrogen cycle. Molecular techniques were used to target a section of the N-fixing gene in a wheat rhizosphere soil. The results suggest an increase in abundance of N-fixing rhizobacteria from which an increase in N fixation could be inferred. Research undertaken on acid soils in NorthEast Victoria showed an increase in ammonium N oxidisers following the application of lime.

Lime also impacts on soil structure by increasing aggregation of soil particles and the creation of a greater diversity of macro and micro pore spaces for improved habitat. Air and water movement through soil is also enhanced. In addition, biological access to food resources can be improved.

Soil Respiration

Soil respiration refers to the production of carbon dioxide when soil organisms respire. This includes respiration of plant roots, the rhizosphere, microbes and fauna.

Soil respiration is a key ecosystem process that releases carbon from the soil in the form of CO_2. CO_2 is acquired from the atmosphere and converted into organic compounds in the process of photosynthesis. Plants use these organic compounds to build structural components or respire them to release energy. When plant respiration occurs below-ground in the roots, it adds to soil respiration. Over time, plant structural components are consumed by heterotrophs. This heterotrophic consumption releases CO_2 and when this CO_2 is released by below-ground organisms, it is considered soil respiration.

The amount of soil respiration that occurs in an ecosystem is controlled by several factors. The temperature, moisture, nutrient content and level of oxygen in the soil can produce extremely disparate rates of respiration. These rates of respiration can be measured in a variety of methods. Other methods can be used to separate the source components, in this case the type of photosynthetic pathway (C3/C4), of the respired plant structures.

Soil respiration rates can be largely affected by human activity. This is because humans have the ability to and have been changing the various controlling factors of soil respiration for numerous years. Global climate change is composed of numerous changing factors including rising atmospheric CO_2, increasing temperature and shifting precipitation patterns. All of these factors can affect the rate of global soil respiration. Increased nitrogen fertilization by humans also has the potential to affect rates over the entire planet.

Soil respiration and its rate across ecosystems is extremely important to understand. This is because soil respiration plays a large role in global carbon cycling as well as other nutrient cycles. The respiration of plant structures releases not only CO_2 but also other nutrients in those structures, such as nitrogen. Soil respiration is also associated with positive feedback with global climate change. Positive feedback is when a change in a system produces response in the same direction of the change. Therefore, soil respiration rates can be affected by climate change and then respond by enhancing climate change.

Sources of Carbon Dioxide in Soil

A portable soil respiration system measuring soil CO_2 flux

All cellular respiration releases energy, water and CO_2 from organic compounds. Any respiration that occurs below-ground is considered soil respiration. Respiration by plant roots, bacteria, fungi and soil animals all release CO_2 in soils, as described below.

Tricarboxylic Acid (TCA) Cycle

The tricarboxylic acid (TCA) cycle – or citric acid cycle – is an important step in cellular respiration. In the TCA cycle, a six carbon sugar is oxidized. This oxidation produces the CO_2 and H_2O from the sugar. Plants, fungi, animals and bacteria all use this cycle to convert organic compounds to energy. This is how the majority of soil respiration occurs at its most basic level. Since the process relies on oxygen to occur, this is referred to as aerobic respiration.

Fermentation

Fermentation is another process in which cells gain energy from organic compounds. In this metabolic pathway, energy is derived from the carbon compound without the use of oxygen. The products of this reaction are carbon dioxide and usually either ethyl alcohol or lactic acid. Due to the lack of oxygen, this pathway is described as anaerobic respiration. This is an important source of CO_2 in soil respiration in waterlogged ecosystems where oxygen is scarce, as in peat bogs and wetlands. However, most CO_2 released from the soil occurs via respiration and one of the most important aspects of below-ground respiration occurs in the plant roots.

Root Respiration

Plants respire some of the carbon compounds which were generated by photosynthesis. When this respiration occurs in roots, it adds to soil respiration. Root respiration accounts for approximately half of all soil respiration. However, these values can range from 10–90% depending on the dominate plant types in an ecosystem and conditions under which the plants are subjected. Thus, the amount of CO_2 produced through root respiration is determined by the root biomass and specific root respiration rates. Directly next to the root is the area known as the rhizosphere, which also plays an important role in soil respiration.

Rhizosphere Respiration

The rhizosphere is a zone immediately next to the root surface with its neighboring soil. In this zone there is a close interaction between the plant and microorganisms. Roots continuously release substances, or exudates, into the soil. These exudates include sugars, amino acids, vitamins, long chain carbohydrates, enzymes and lysates which are released when roots cells break. The amount of carbon lost as exudates varies considerably between plant species. It has been demonstrated that up to 20% of carbon acquired by photosynthesis is released into the soil as root exudates. These exudates are decomposed primarily by bacteria. These bacteria will respire the carbon compounds through the TCA cycle; however, fermentation is also present. This is due to the lack of oxygen due to greater oxygen consumption by the root as compared to the bulk soil, soil at a greater distance from the root. Another important organism in the rhizosphere are root-infecting fungi or mycorrhizae. These fungi increase the surface area of the plant root and allow the root to encounter and acquire a greater amount of soil nutrients necessary for plant growth. In return for this benefit, the plant will transfer sugars to the fungi. The fungi will respire these sugars for energy thereby increasing soil respiration. Fungi, along with bacteria and soil animals, also play a large role in the decomposition of litter and soil organic matter.

Soil Animals

Soil animals graze on populations of bacteria and fungi as well as ingest and break up litter to increase soil respiration. Microfauna are made up of the smallest soil animals. These include nematodes and mites. This group specializes on soil bacteria and fungi. By ingesting these organisms, carbon that was initially in plant organic compounds and was incorporated into bacterial and fungal structures will now be respired by the soil animal. Mesofauna are soil animals from 0.1 to 2 millimeters (0.0039 to 0.0787 in) in length and will ingest soil litter. The fecal material will hold a greater amount of moisture and have a greater surface area. This will allow for new attack by microorganisms and a greater amount of soil respiration. Macrofauna are organisms from 2 to 20 millimeters (0.079 to 0.787 in), such as earthworms and termites. Most macrofauna fragment litter, thereby exposing a greater amount of area to microbial attack. Other macrofauna burrow or ingest litter, reducing soil bulk density, breaking up soil aggregates and increasing soil aeration and the infiltration of water.

Regulation of Soil Respiration

Regulation of CO_2 production in soil is due to various abiotic, or non-living, factors. Temperature, soil moisture and nitrogen all contribute to the rate of respiration in soil.

Temperature

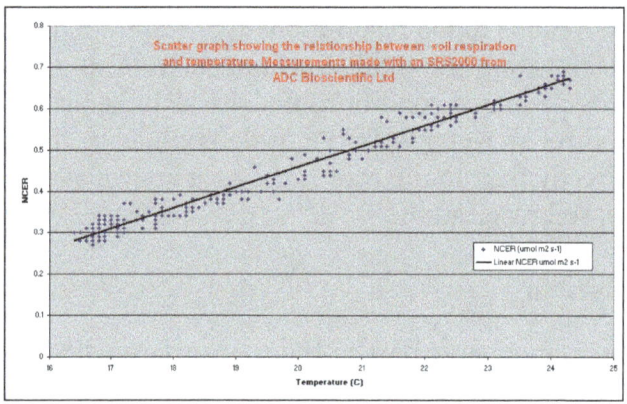

Graph showing soil respiration vs. soil temperature

Temperature affects almost all aspects of respiration processes. Temperature will increase respiration exponentially to a maximum, at which point respiration will decrease to zero when enzymatic activity is interrupted. Root respiration increases exponentially with temperature in its low range when the respiration rate is limited mostly by the TCA cycle. At higher temperatures the transport of sugars and the products of metabolism become the limiting factor. At temperatures over 35 °C (95 °F), root respiration begins to shut down completely. Microorganisms are divided into three temperature groups; cryophiles, mesophiles and thermophiles. Cryophiles function optimally at temperatures below 20 °C (68 °F), mesophiles function best at temperatures between 20 and 40 °C (104 °F) and thermophiles function optimally at over 40 °C (104 °F). In natural soils many different cohorts, or groups of microorganisms exist. These cohorts will all function best at different conditions, so respiration may occur over a very broad range. Temperature increases lead to greater rates of soil respiration until high values retard microbial function, this is the same pattern that is seen with soil moisture levels.

Soil Moisture

Soil moisture is another important factor influencing soil respiration. Soil respiration is low in dry conditions and increases to a maximum at intermediate moisture levels until it begins to decrease when moisture content excludes oxygen. This allows anaerobic conditions to prevail and depress aerobic microbial activity. Studies have shown that soil moisture only limits respiration at the lowest and highest conditions with a large plateau existing at intermediate soil moisture levels for most ecosystems. Many microorganisms possess strategies for growth and survival under low soil moisture conditions. Under high soil moisture conditions, many bacteria take in too much water causing their cell membrane to lyse, or break. This can decrease the rate of soil respiration temporarily, but the lysis of bacteria causes for a spike in resources for many other bacteria. This rapid increase in available labile substrates causes short-term enhanced soil respiration. Root respiration will increase with increasing soil moisture, especially in dry ecosystems; however, individual species' root respiration response to soil moisture will vary widely from species to species depending on life history traits. Upper levels of soil moisture will depress root respiration by restricting access to atmospheric oxygen. With the exception of wetland plants, which have developed specific mechanisms for root aeration, most plants are not adapted to wetland soil environments with low oxygen. The respiration dampening effect of elevated soil moisture is amplified when soil respiration also lowers soil redox through bioelectrogenesis. Soil-based microbial fuel cells are becoming popular educational tools for science classrooms.

Nitrogen

Nitrogen directly affects soil respiration in several ways. Nitrogen must be taken in by roots in order to promote plant growth and life. Most available nitrogen is in the form of NO_3^-, which costs 0.4 units of CO_2 to enter the root because energy must be used to move it up a concentration gradient. Once inside the root the NO_3^- must be reduced to NH_3. This step requires more energy, which equals 2 units of CO_2 per molecule reduced. In plants with bacterial symbionts, which fix atmospheric nitrogen, the energetic cost to the plant to acquire one molecule of NH_3 from atmospheric N_2 is 2.36 CO_2. It is essential that plants uptake nitrogen from the soil or rely on symbionts to fix it from the atmosphere in order to assure growth, reproduction and long-term survival.

Another way nitrogen affects soil respiration is through litter decomposition. High nitrogen litter is considered high quality and is more readily decomposed by microorganisms than low quality litter. Degradation of cellulose, a tough plant structural compound, is also a nitrogen limited process and will increase with the addition of nitrogen to litter.

Methods of Measurement

Different methods exist for the measurement of soil respiration rate and the determination of sources. The most common methods include the use of long-term stand alone soil flux systems for measurement at one location at different times; survey soil respiration systems for measurement of different locations and at different times; and the use of stable isotope ratios.

Long-term Stand-alone Soil Flux Systems for Measurement at one Location over Time

These systems measure at one location over long periods of time. Since they only measure at one location, it is common to use multiple stations to reduce measuring error caused by soil variability over small distances. Soil variability may be tested with survey soil respiration instruments.

The long-term instruments are designed to expose the measuring site to ambient conditions as much as is possible between measurements.

An Automated Soil CO_2 Exchange System

Types of Long-term Stand-alone Instruments

Closed Mode Systems

Closed mode systems, by definition, are closed without a vent. When making measurements, closed systems take a CO_2 measurement of ambient air just after the soil sample chamber is sealed. This becomes the CO_2 reference. The system continues to take measurements at regular intervals over the next few minutes until a final CO_2 concentration is recorded at a predetermined end time over time. Closed systems have the advantages that they are used by more researchers, and they provide wind proof results. Since there is no vent, wind is not an issue. When using a closed mode system, it is important that the experimental design include a maximum measuring time for two reasons: First; if the measuring time is too long, an unacceptable positive partial CO_2 pressure can build up in the chamber limiting the amount of additional CO_2 from entering the chamber, causing a measuring artifact. The second reason is that if a chamber that is closed for too long, it can modify ambient conditions and cause measuring artifacts. Finding the optimal measuring time range is important. Tests on the soil in question can confirm the best time solution, and eliminate the partial CO_2 pressure buildup from being a significant issue.

Both individual assay information and diurnal CO_2 respiration measuring information is accessible. It is also common for such systems to also measure soil temperature, soil moisture and PAR (photosynthetically active radiation). These variables are normally recorded in the measuring file along with CO_2 values.

For determination of soil respiration and the slope of CO_2 increase, researchers have used linear regression analysis, the Pedersen algorithm, and exponential regression. There are more published references for linear regression analysis; however, the Pedersen algorithm and exponential regression analysis methods also have their following. Some systems offer a choice of mathematical methods.

When using linear regression, multiple data points are graphed and the points can be fitted with a linear regression equation, which will provide a slope. This slope can provide the rate of soil respiration with the equation $F = bV / A$, where F is the rate of soil respiration, b is the slope, V is the volume of the chamber and A is the surface area of the soil covered by the chamber. It is important that the measurement is not allowed to run over a longer period of time as the increase in CO_2

concentration in the chamber will also increase the concentration of CO_2 in the porous top layer of the soil profile. This increase in concentration will cause an underestimation of soil respiration rate due to the additional CO_2 being stored within the soil.

Open Mode Systems

Open mode systems are designed to find soil flux rates when measuring chamber equilibrium has been reached. Air flows through the chamber before the chamber is closed and sealed. This purges any non-ambient CO_2 levels from the chamber before measurement. After the chamber is closed, fresh air is pumped into the chamber at a controlled and programmable flow rate. This mixes with the CO_2 from the soil, and after a time, equilibrium is reached. The researcher specifies the equilibrium point as the difference in CO_2 measurements between successive readings, in an elapsed time. During the assay, the rate of change slowly reduces until it meets the customer's rate of change criteria, or the maximum selected time for the assay. Soil flux or rate of change is then determined once equilibrium conditions are reached within the chamber. Chamber flow rates and times are programmable, accurately measured, and used in calculations. These systems have vents that are designed to prevent a possible unacceptable buildup of partial CO_2 pressure discussed under closed mode systems. Since the air movement inside the chamber might cause increased chamber pressure, or external winds may produce reduced chamber pressure, a vent is provided that is designed to be as wind proof as possible.

Open systems are also not as sensitive to soil structure variation, or to boundary layer resistance issues at the soil surface. Air flow in the chamber at the soil surface is designed to minimize boundary layer resistance phenomena.

Hybrid Mode Systems

A hybrid system also exists. It has a vent that is designed to be as wind proof as possible, and prevent possible unacceptable partial CO_2 pressure buildup, but is designed to operate like a closed mode design system in other regards.

Survey Soil Respiration Systems – for Testing the Variation of CO_2 Respiration at Different Locations and at Different Times.

Measuring spatial variability of soil respiration in the field

These are either open or closed mode instruments that are portable or semi-portable. They measure CO_2 soil respiration variability at different locations and at different times. With this type of instrument, soil collars that can be connected to the survey measuring instrument are inserted into the ground and the soil is allowed to stabilize for a period of time. The insertion of the soil collar temporarily disturbs the soil, creating measuring artifacts. For this reason, it is common to have several soil collars inserted at different locations. Soil collars are inserted far enough to limit lateral diffusion of CO_2. After soil stabilization, the researcher then moves from one collar to another according to experimental design to measure soil respiration.

Survey soil respiration systems can also be used to determine the number of long-term stand-alone temporal instruments that are required to achieve an acceptable level of error. Different locations may require different numbers of long-term stand-alone units due to greater or lesser soil respiration variability.

Isotope Methods

Plants acquire CO_2 and produce organic compounds with the use of one of three photosynthetic pathways. The two most prevalent pathways are the C_3 and C_4 processes. C_3 plants are best adapted to cool and wet conditions while C_4 plants do well in hot and dry ecosystems. Due to the different photosynthetic enzymes between the two pathways, different carbon isotopes are acquired preferentially. Isotopes are the same element that differ in the number of neutrons, thereby making one isotope heavier than the other. The two stable carbon isotopes are ^{12}C and ^{13}C. The C_3 pathway will discriminate against the heavier isotope more than the C_4 pathway. This will make the plant structures produced from C_4 plants more enriched in the heavier isotope and therefore root exudates and litter from these plants will also be more enriched. When the carbon in these structures is respired, the CO_2 will show a similar ratio of the two isotopes. Researchers will grow a C_4 plant on soil that was previously occupied by a C_3 plant or vice versa. By taking soil respiration measurements and analyzing the isotopic ratios of the CO_2 it can be determined whether the soil respiration is mostly old versus recently formed carbon. For example, maize, a C_4 plant, was grown on soil where spring wheat, a C_3 plant, was previously grown. The results showed respiration of C_3 SOM in the first 40 days, with a gradual linear increase in heavy isotope enrichment until day 70. The days after 70 showed a slowing enrichment to a peak at day 100. By analyzing stable carbon isotope data it is possible to determine the source components of respired SOM that was produced by different photosynthetic pathways.

Responses to Human Disturbance

Throughout the past 160 years, humans have changed land use and industrial practices, which have altered the climate and global biogeochemical cycles. These changes have affected the rate of soil respiration around the planet.

Elevated Carbon Dioxide

Since the Industrial Revolution, humans have emitted vast amounts of CO_2 into the atmosphere. These emissions have increased greatly over time and have increased global atmospheric CO_2 levels to their highest in over 750,000 years. Soil respiration increases when ecosystems are exposed to elevated levels of CO_2. Numerous free air CO_2 enrichment (FACE) studies have been conducted to test soil respiration under predicted future elevated CO_2 conditions. Recent FACE studies have shown large increases in soil respiration due to increased root biomass and microbial activity. Soil respiration has been found to increase up to 40.6% in a sweetgum forest in Tennessee and poplar forests in Wisconsin under elevated CO_2 conditions. It is extremely likely that CO_2 levels will exceed those used in these FACE experiments by the middle of this century due to increased human use of fossil fuels and land use practices.

Climate Warming

Due to the increase in temperature of the soil, CO_2 levels in our atmosphere increase, and as such the mean average temperature of the Earth is rising. This is due to human activities such as forest clearing, soil denuding, and developments that destroy autotrophic processes. With the loss of photosynthetic plants covering and cooling the surface of the soil, the infrared energy penetrates the soil heating it up and causing a rise in heterotrophic bacteria. Heterotrophs in the soil quickly degrade the organic matter and soil structure crumbles, thus it dissolves into streams and rivers into the sea. Much of the organic matter swept away in floods caused by forest clearing goes into estuaries, wetlands and eventually into the open ocean. Increased turbidity of surface waters causes biological oxygen demand and more autotrophic organisms die. Carbon dioxide levels rise with increased respiration of soil bacteria after temperatures rise due to loss of soil cover.

As mentioned earlier, temperature greatly affects the rate of soil respiration. This may have the most drastic influence in the Arctic. Large stores of carbon are locked in the frozen permafrost. With an increase in temperature, this permafrost is melting and aerobic conditions are beginning to prevail, thereby greatly increasing the rate of respiration in that ecosystem.

Changes in Precipitation

Due to the shifting patterns of temperature and changing oceanic conditions, precipitation patterns are expected to change in location, frequency and intensity. Larger and more frequent storms are expected when oceans can transfer more energy to the forming storm systems. This may have the greatest impact on xeric, or arid, ecosystems. It has been shown that soil respiration in arid ecosystems shows dynamic changes within a raining cycle. The rate of respiration in dry soil usually bursts to a very high level after rainfall and then gradually decreases as the soil dries. With an increase in rainfall frequency and intensity over area without previous extensive rainfall, a dramatic increase in soil respiration can be inferred.

Nitrogen Fertilization

Since the onset of the Green Revolution in the middle of the last century, vast amounts of nitrogen fertilizers have been produced and introduced to almost all agricultural systems. This has led to increases in plant available nitrogen in ecosystems around the world due to agricultural runoff and wind-driven fertilization. Nitrogen can have a significant positive effect on the level and rate of soil respiration. Increases in soil nitrogen have been found to increase plant dark respiration, stimulate specific rates of root respiration and increase total root biomass. This is because high nitrogen rates are associated with high plant growth rates. High plant growth rates will lead to the increased respiration and biomass found in the study. With this increase in productivity, an increase in soil activities and therefore respiration can be assured.

Importance

Soil respiration plays a significant role in the global carbon and nutrient cycles as well as being a driver for changes in climate. These roles are important to our understanding of the natural world and human preservation.

Global Carbon Cycling

Soil respiration plays a critical role in the regulation of carbon cycling at the ecosystem level and at global scales. Each year approximately 120 petagrams (Pg) of carbon are taken up by land plants and a similar amount is released to the atmosphere through ecosystem respiration. The global soils contain up to 3150 Pg of carbon, of which 450 Pg exist in wetlands and 400 Pg in permanently frozen soils. The soils contain more than four times the carbon as the atmosphere. Researchers have estimated that soil respiration accounts for 77 Pg of carbon released to the atmosphere each year. This level of release is one order of magnitude greater than the carbon release due to anthropogenic sources (6 Pg per year) such as fossil fuel burning. Thus, a small change in soil respiration can seriously alter the balance of atmosphere CO_2 concentration versus soil carbon stores. Much like soil respiration can play a significant role in the global carbon cycle, it can also regulate global nutrient cycling.

Nutrient Cycling

A major component of soil respiration is from the decomposition of litter which releases CO_2 to the environment while simultaneously immobilizing or mineralizing nutrients. During decomposition, nutrients such as nitrogen are immobilized by microbes for their own growth. As these microbes are ingested or die, nitrogen is added to the soil. Nitrogen is also mineralized from the degradation of proteins and nucleic acids in litter. This mineralized nitrogen is also added to the soil. Due to these processes, the rate of nitrogen added to the soil is coupled with rates of microbial respiration. Studies have shown that rates of soil respiration were associated with rates of microbial turnover and nitrogen mineralization. Alterations of the global cycles can further act to change the climate of the planet.

Climate Change

As stated earlier, the CO_2 released by soil respiration is a greenhouse gas that will continue to trap energy and increase the global mean temperature if concentrations continue to rise. As global temperature rises, so will the rate of soil respiration across the globe thereby leading to a higher concentration of CO_2 in the atmosphere, again leading to higher global temperatures. This is an example of a positive feedback loop. It is estimated that a rise in temperature by 2° Celsius will lead to an additional release of 10 Pg carbon per year to the atmosphere from soil respiration. This is a larger amount than current anthropogenic carbon emissions. There also exists a possibility that this increase in temperature will release carbon stored in permanently frozen soils, which are now melting. Climate models have suggested that this positive feedback between soil respiration and temperature will lead to a decrease in soil stored carbon by the middle of the 21st century.

Biological Soil Crust

Biological soil crusts are also known as cryptogamic, microbiotic, cryptobiotic, and microphytic crusts, leading to some confusion. The names are all meant to indicate common features of the organisms that compose the crusts. The most inclusive term is probably biological soil crust, as this distinguishes them from physical crusts without limiting the crust components to plants. Whatever name used, there remains an important distinction between these formations and physical or chemical crusts.

Biological soil crusts are formed by living organisms and their by-products, creating a crust of soil particles bound together by organic materials.

Chemical and physical crusts are inorganic features such as a salt crust or platy surface crust, often formed by trampling.

Structure and Formation

Mature crusts of the Colorado Plateau

Undisturbed crusts on the Colorado Plateau are usually darker than the disturbed soils

Crusts generally cover all soil spaces not occupied by vascular plants, and may be 70% or more of the living cover

Crusts are formed by living organisms and their by-products, creating a surface crust of soil particles bound together by organic materials. Aboveground crust thickness can reach up to 10 cm. The general appearance of the crusts in terms of color, surface topography, and surficial coverage varies. Mature crusts of the Great Basin and Colorado Plateau are usually darker than the surrounding soil. This color is due in part to the density of the organisms and to the often dark color of the cyanobacteria, lichens, and mosses. Crusts generally cover all soil spaces not occupied by vascular plants, and may be 70% or more of the living cover.

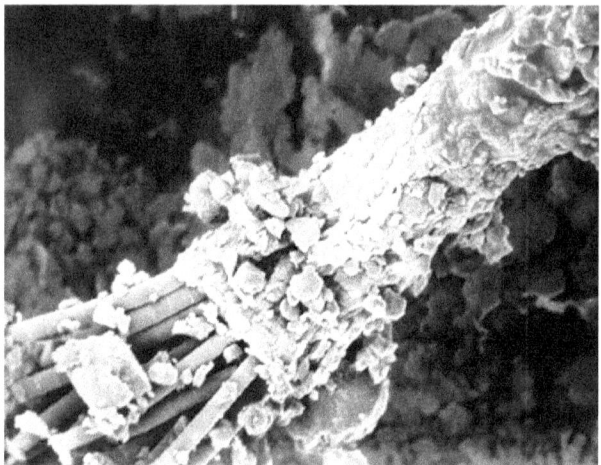

Filamentous cyanobacteria migrating out of their sheaths, x 950.

These crusts are characterized by their marked increase in surface topography, often referred to as pinnacles or pedicles. The process of creating surface topography, or pinnacling, is due largely to the presence of filamentous cyanobacteria and green algae. These organisms swell when wet, migrating out of their sheaths. After each migration new sheath material is exuded, thus extending sheath length. Repeated swelling leaves a complex network of empty sheath material that maintains soil structure after the organisms have dehydrated and decreased in size. Frost heaving, subsequent uneven erosion, and lack of surface plant roots results in high pedicles. In in warmer regions such as the the Sonoran, Mojave, and Chihuahuan deserts, lack of frost heaving has been used to explain the absence of pinnacles. In northern deserts, where most rain falls in the winter and surface plant roots are plentiful, crusts are generally rolling or smooth.

Mature crust in the Mojave desert.

Species Composition

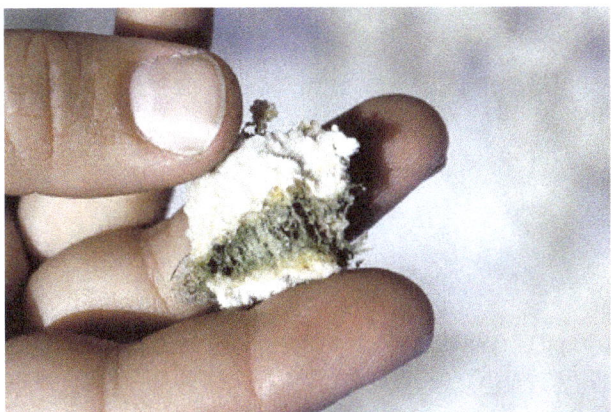

Soil cyanobacteria can either be on, or beneath, the soil surface.

Crusts are predominantly composed of cyanobacteria (formerly called blue-green algae), green and brown algae, mosses, and lichens. Liverworts, fungi, and bacteria can also be important components.

Soil mosses and lichens found on crust.

In the Great Basin and the Colorado Plateau, Microcolues vaginatus (a cyanobacteria) composes the vast majority of the crust structure. Lichens of the genera Collema spp. and mosses from the genera Tortula spp. are common. In hot deserts, such as the Sonoran, other cyanobacteria are more common. Some more acidic soils are dominated by green algae. Shifts between green algal and cyanobacterial dominance have been attributed to changes in pH, with the decreasing alkalinity favoring green algae. More stable crusts are dominated by lichens and/or mosses. The organism that dominates the crust is partly determined by microclimate and may also represent different successional stages in crust development.

Ecological Functions

Crusts play an important role in the environment. Because they are concentrated in the top 1 to 4 mm of soil, they primarily effect processes that occur at the land surface or soil-air interface. These include soil stability and erosion, atmospheric nitrogen-fixation, nutrient contributions to plants, soil-plant-water relations, infiltration, seedling germination, and plant growth.

Soil Stability

Scanning electron micrograph of cyanobacterial sheath material, sticking to sand grains, x 90.

Scanning electron micrograph of cyanobacterial sheath material, holding sand grains together, x 100.

Cyanobacterial sheath material holding soil particles together.

Steep slope held in place by crusts.

Crust-forming cyanobacteria have filamentous growth forms that bind soil particles. These filaments exude sticky polysaccharide sheaths around their cells that aid in soil aggregation by cementing particles together. Fungi, both free-living and as a part of lichens, contribute to soil stability by binding soil particles with hyphae. Lichens and mosses assist in soil stability by binding particles with rhizines/rhizoids, increasing resistance to wind and water erosion. The increased surface topography of some crusts, along with increased aggregate stability, further improves resistance to wind and water erosion.

Soils less than 6" deep, held in place by crusts.

Water Infiltration

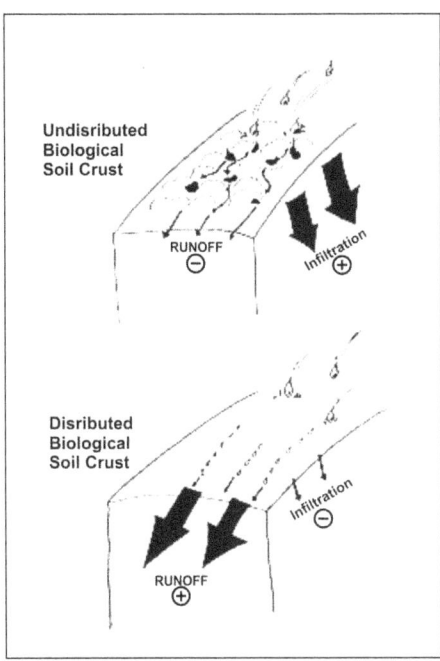

Top figure: Roughened soil surfaces slow the velocity of water runoff, resulting in longer residence times.
Longer residence times increase water infiltration into the soil. Bottom figure: Smooth surfaces
do not slow runoff water, and thus infiltration is decreased relative to rough surfaces.

Crusts can alter water infiltration. Studies where crusts greatly increase surface roughness generally have increased infiltration. Where crusts do not significantly increase surface roughness, infiltration is generally reduced due to the presence of cyanobacterial filaments. Differences in findings are therefore site-specific and also related to soil texture and chemical properties of the soil.

Effects on Plant Germination and Growth

Studies investigating the role of crusts in plant germination have had varied results. Increased surface relief is believed to provide safe sites for seeds, while darker surface color increases soil temperatures to those required for germination earlier in the season, coinciding with spring water availability.

Large-seeded plants often require burial for germination. Native seeds have self-drilling mechanisms or are cached by rodents. However, soil crusts reduce soil movement, and this may limit passive burial and germination of large-seeded exotic plants like Bromus tectorum (cheatgrass).

Studies of how crusts effect plant health are clear-cut. Many studies have shown increases in survival and nutrient content in crust-covered environments as opposed to bare soil. Nutrients shown to increase in plant tissues grown in the presence of crusts are nitrogen, phosphorus, potassium, iron, calcium, magnesium, and manganese. Some of the plants benefited by crust presence include Festuca octoflora (sixweeks fescue), Mentzelia multiflora (desert blazing star), Arabis fecunda (rock-cress), Kochia prostrata (prostrate summercypress), Linum perenne (blue flax), Lepidium montanum (mountain peppergrass), and Sphaeralcea coccinea (scarlet globemallow).

Response to Disturbance

Compressional disturbances crushes crusts, leaving soils
underneath them unprotected from wind or water erosion.

Crusts are well adapted to severe growing conditions, but poorly adapted to compressional disturbances. Domestic livestock grazing, and more recently, recreational activities (hiking, biking, and off-road driving) and military activities place a heavy toll on the integrity of the crusts. Disruption of the crusts brings decreased organism diversity, soil nutrients, stability, and organic matter.

Removal of crust can lead to large soil losses, as seen here in the Channel Islands.

Direct damage to crusts usually comes in the form of trampling by humans and livestock or vehicles driving off of roads. Compressional disturbances break sheaths and filaments and drastically reduces the ability of the soil organisms to function, particularly in providing nitrogen and soil stability. Changes in plant composition are often used as indicators of range health. This indicator may not be sensitive enough to warn of damage to microbiotic crusts. Studies of trampling disturbance have noted that losses of moss cover, lichen cover, and cyanobacterial presence can be severe (1/10, 1/3, and 1/2 respectively), runoff can increase by half, and the rate of soil loss can increase six times without apparent damage to vegetation. Disturbance to soil surfaces in arid regions can lead to large soil losses.

Disturbance to soil surfaces in arid regions can result in large amounts
of soil loss. Here, soil levels are now several feet below what
they were when this tree was alive.

Other disturbance impacts are indirect. Several native rangeland shrubs (Artemisia tridentata, Atriplex confertifolia, and Ceratoides lanata) may have allelopathic effects on the nitrogen-fixing capabilities of crusts, potentially lowering nitrogen fixation by 80 percent. Actions that increase the shrub component, such as excessive grazing, can have an unexpected impact on crust functioning.

Another indirect disturbance occurs through crust burial. When the integrity of the crust is broken through trampling or other means, the soil is more susceptible to wind and water erosion. This soil can be moved long distances, covering intact crusts. Crusts tolerate shallow burial by extending sheaths to the surface to begin photosynthesis again. Deeper burial by eroded sediment will kill crusts.

Blowing or washing sand buries crusts, killing them.

Fire is a common occurrence in many regions where microbiotic crusts grow. Investigations show that fires can cause severe damage, but that recovery is possible. The degree to which crusts are damaged by fires apparently depends on the intensity of the fire. Low-intensity fires do not remove all of the crust structure, which allows for regrowth without significant soil loss. Shrub presence (particularly sagebrush) increases the intensity of the fire, decreasing the likelihood of early vegetative or crust recovery.

Full recovery of crust from disturbance is a slow process, particularly for mosses and lichens. There are means to facilitate recovery. Allowing the cyanobacterial and green algae component to recover will give the appearance of a healthy crust. This visual recovery can be complete in as little as 1 to 5 years, given average climate conditions. However, recovering crust thickness can take up to 50 years, and mosses and lichens can take up to 250 years to recover. Limiting the size of the disturbed area also increases the rate of recovery, provided that there is a nearby source of inoculum.

Cyanobacterial thickness 5 years after disturbance.
Thickness increases about 1 mm/year.

Main Components of Soil Crusts

Cyanobacteria (also called "blue-green algae") are often the first soil crust organisms to colonize an area after a disturbance. These primitive bacteria are photosynthetic and can capture atmospheric nitrogen into a form that is available to vascular plants. Thin filaments of cyanobacteria secrete sticky substances that bind soil particles together. One of the most common cyanobacterium in the Colorado Plateau and Great Basin Desert (and also worldwide) is Microcoleus vaginatus, although other species may be more common in the hot deserts of the Southwestern United States. Green algae are light green to black photosynthetic organisms occurring as single cells or colonies. In biological soil crusts, they are found on or just below the soil surface.

Green algae dry out and become dormant during dry times, but they "wake up" with even small amount of moisture. Unlike their aquatic counterparts, green algae in crusts are well adapted to living and reproducing in dry desert environments.

Lichen are a part of the living community that make up biological soil crust

Fungi in biological soil crusts usually occur as free-living organisms, but they can also form symbiotic relationships with plant roots. Free-living fungi function as decomposers, feeding on organic material such as leaf litter, and contribute to the cycling of nutrients in the soil crust. Like the cyanobacteria, fungal filaments secrete substances that help bind soil particles together and increase soil stability.

Bryophytes are small, non-vascular plants known as mosses and liverworts, with mosses being more common in soil crust communities.

Lichens are symbiotic systems involving a fungal partner and photosynthetic alga or cyanobacterium. The alga or cyanobacterium provide the fungal partner with food (carbohydrates), while the fungus provides a suitable environment by effectively regulating moisture and sunlight. Lichens come in a wide variety of shapes, sizes, and colors.

Soil System

Soils are major components of the world's ecosystems. Soil forms the Earth's atmosphere, lithosphere (rocks), biosphere (living matter) and hydrosphere (water). Soil is what forms the outermost layer of the Earth's surface, and comprise weathered bedrock (regolith), organic matter (both dead and alive), air and water.

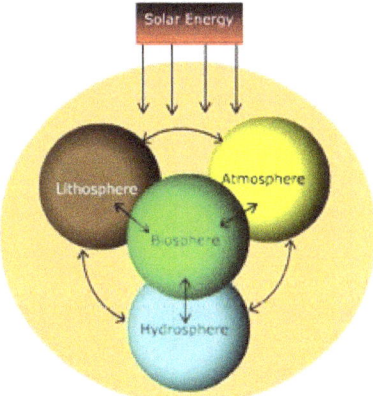

The soil interacts with the atmosphere, lithosphere, biosphere and hydrosphere:

- The water cycle moves through the soil by infiltration and water may evaporate from the surface.

- The atmosphere may contain particulate matter that is deposited on the soils and particles may blow up into the atmosphere.

- Rocks in the lithosphere weather to form soils, and soils at depth and pressure may form rocks.

- Plants in the biosphere may extract nutrients from the soils and dead plants may end up forming parts of the soil.

Soils are important to humans in many ways:

- Soil is the medium for plant growth, which most of foods for humans are grown in.

- Soil stores freshwater, 0.005% of world's freshwater.

- Soil filters materials added to the soil, keeping quality water.

- Recycling of nutrients takes place in the soil when dead organic matter is broken down.

- Soil is the habitat for billions of micro-organisms, as well as other larger animals.

- Soil provides raw material in the forms of peat, clay, sands, gravel and minerals.

Soil Horizon

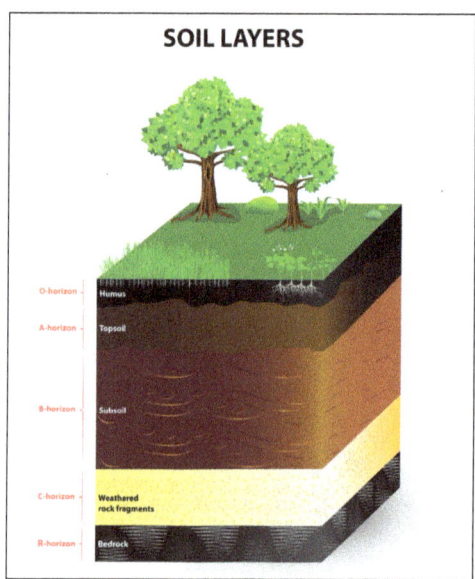

- O) Organic matter: Litter layer of plant residues in relatively undecomposed form.

- A) Surface Soil: Layer of mineral soil with most organic matter accumulation and soil life. This layer eluviates (is depleted of) iron, clay and calcium, organic compounds, and other soluble constituents. When eluviation is pronounced, a lighter colored "E" subsurface soil horizon is apparent at the base of the "A" horizon. A-horizons may also be the result of a combination of soil bioturbation and surface processes that separates fine particles from biologically mounded topsoil. In this case, the A-horizon is regarded as a "biomantle".

- B) Subsoil: This layer accumulates iron, clay, aluminum and organic compounds, a process referred to as illuviation.

- C) Parent Rock: Layer of large unbroken rocks. This layer may accumulate the more soluble compounds.

- R) Bedrock: The parent material in bedrock landscapes. This layer denotes the layer of partially weathered bedrock at the base of the soil profile. Unlike the above layers, R horizons largely comprise continuous masses of hard rock that cannot be excavated by hand. Soils formed *in situ* will exhibit strong similarities to this bedrock layer. These areas of bedrock are under 50 feet of the other profiles.

Soil system storages include organic matter, organisms, nutrients, minerals, air and water.

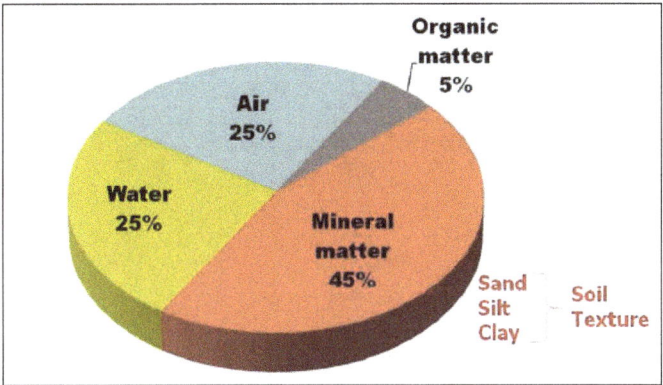

Soil has matter in all three states:

- Organic and inorganic matter form the solid state.

- Soil water(from precipitation, groundwater and seepage) form the liquid state.

- Soil atmosphere forms the gaseous state.

Transfers of material within the soil, including biological mixing and leaching (minerals dissolved in water moving through soil), contribute to the organization of the soil.

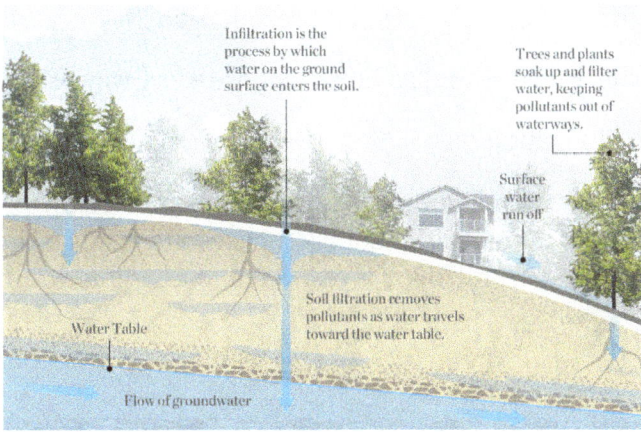

Translocation involves the movement of soil-forming materials through the developing soil profile. Translocation occurs by water running through the soil transferring materials from upper to lower portions of the profile. Burrowing animals like earth worms, ants, etc., move soil materials within the profile. Burrowing animals create passage ways through which air and water can travel promoting soil development.

There are inputs of organic material including leaf litter and inorganic matter from parent material, precipitation and energy. Outputs include uptake by plants and soil erosion.

INPUTS	OUTPUTS
Soil	
Fertilizer	Crop Removal
N Fixation	Leaching Loss
Plant & Animal Residues	Denitrification
Precipitation	Volatilization
Available pool	
Mineralization	Immobilization

Inputs

1. Weathering

    Rock weathering is one of the most important long-term sources for nutrients. However, this process adds nutrients to ecosystems in relatively small quantities over long periods of time. Important nutrients released by weathering include:

    Calcium, magnesium, potassium, sodium, silicon, iron, aluminum, and phosphorus.

2. Atmospheric Input

    *   Large quantities of nutrients are added to ecosystems from the atmosphere. This addition is done either through precipitation or by a number of biological processes.

    *   Carbon - absorbed by way of photosynthesis.

    *   Nitrogen - produced by lightning and precipitation.

    *   Sulfur, chloride, calcium, and sodium - deposited by way of precipitation.

3. Biological Nitrogen Fixation

    Biological nitrogen fixation is a biochemical process where nitrogen gas from the atmosphere is chemically combined into more complex solid forms by metabolic reactions in an organism. This ability to fix nitrogen is restricted to a symbiotic associations with legumes and other microorganisms.

Outputs

1. Erosion

    Soil erosion is probably the most import means of nutrient loss to ecosystems. Erosion is very active in agricultural and forestry systems, where cultivation, grazing, and clear-cutting leaves the soil bare and unprotected. When unprotected, the surface of the soil is easily transported by wind and moving water. The top most layers of a soil, which have an abundance of nutrient rich organic matter, are the major storehouse for soil nutrients like phosphorus, potassium, and nitrogen.

2. Leaching

   Leaching occurs when water flowing vertically through the soil transports nutrients in solution downward in the soil profile. Many of these nutrients can be completely lost from the soil profile if carried into groundwater and then horizontally transported into rivers, lakes, or oceans. Leaching losses are, generally, highest in disturbed ecosystems. In undisturbed ecosystems, efficient nutrient cycling limits the amount of nutrients available for this process.

3. Gaseous Losses

   High losses of nutrients can also occur when specific environmental conditions promote the export of nutrients in a gaseous form. When the soil is wet and anaerobic, many compounds are chemically reduced to a gas from solid forms in the soil. This is especially true of soil nitrogen.

Transformations include decomposition, weathering and nutrient cycling.

The transformation and movement of materials within soil organic matter pools is a dynamic process influenced by climate, soil type, vegetation and soil organisms. All these factors operate within a hierarchical spatial scale. Soil organisms are responsible for the decay and cycling of both macronutrients and micronutrients, and their activity affects the structure, tilth and productivity of the soil.

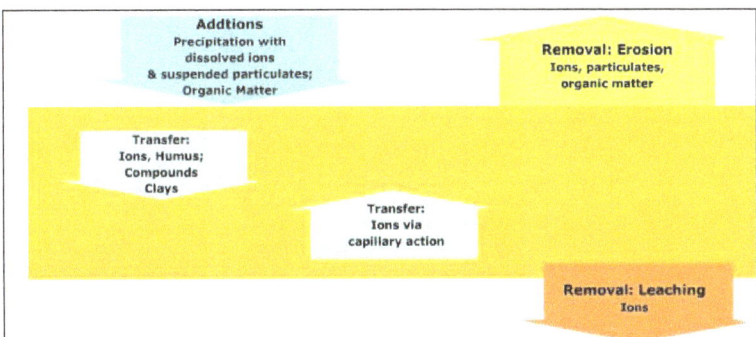

The structure and properties of sand, clay and loam soils differ in many ways, including mineral and nutrient content, drainage, water-holding capacity, air spaces, biota and potential to hold organic matter. Each of these variables is linked to the ability of the soil to promote primary productivity

Soil structure affects aeration, water-holding capacity, drainage, and penetration by roots and seedlings, among other things. Soil structure refers to the arrangement of soil particles into aggregates (or peds) and the distribution of pores in between. It is not a stable property and is greatly influenced by soil management practices.

Primary productivity of soils depend on:

- Mineral content

- Drainage

- Water-holding capacity

- Air spaces

- Biota

- Potential to hold organic matter.

	Water retention and availability	Nutrient storage capacity	Air Space	Primary Production
Clay	Sticky and easily waterlogged	High	Low	Meduim/Low
Sand	Fast draining soils that dry easily	Low	High	Low
Loam	High to medium	Medium	Medium	Medium

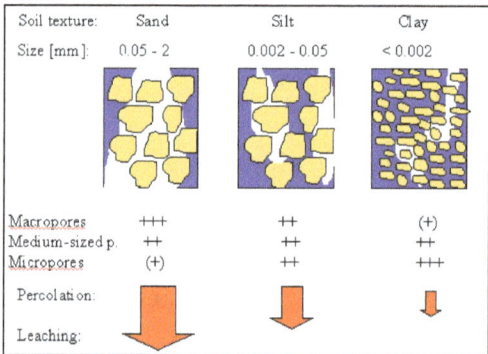

Soil texture and percolation image from soils.

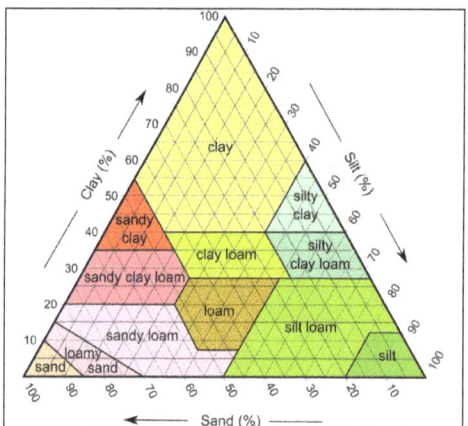

Consider mineral content, drainage, water-holding capacity, air spaces, biota and potential to hold organic matter, and link these to primary productivity.

Soil structure depends on:

- Soil texture (the amount of sand and clay).

- Dead organic matter.

- Earthworm activity.

For optimum structure, variety of pure sizes are required to allow root prevention, free drainage and water storage. Pore spaces over 0.1 mm allow roots growth, oxygen diffusion and water movement where as pore spaces below 0.5 mm help store water.

Application and Skills

Outline the transfers, transformations, inputs, outputs, flows and storages within soil systems.

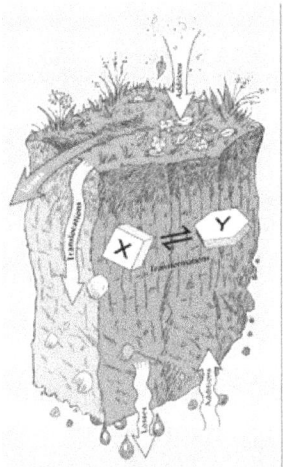

There are four basic processes that occur in the formation of soils:

- Inputs - physical movement of material within soil.

- Outputs - occur both from the surface and from the deep subsoil. Water lost by evapotranspiration.

- Translocations - translocation of materials within the soil profile is primarily due to gradients in water potential and chemical concentrations within the soil pores.

- Transformations - change of some soil constituent without any physical displacement.

The two driving forces for these processes are climate (temperature and precipitation) and organisms, (plants and animals). Parent material is usually a rather passive factor in affecting soil processes because parent materials are inherited from the geologic world. Topography (or relief) is also rather passive in affecting soil processes, mainly by modifying the climatic influences of temperature and precipitation.

Explain how soil can be viewed as an ecosystem.

Soil is the link between the air, water, rocks, and organisms, and is responsible for many different functions in the natural world that we call ecosystem services. These soil functions include: air quality and composition, temperature regulation, carbon and nutrient cycling, water cycling and quality, natural "waste" (decomposition) treatment and recycling, and habitat for most living things and their food. We could not survive without these soil functions.

Billions of organisms inhabit the upper layers of the soil, where they break down dead organic matter, releasing the nutrients necessary for plant growth. The micro-organisms include bacteria, actinomycetes, algae and fungi. Macro-organisms include earthworms and arthropods such as insects, mites and millipedes. Each group plays a role in the soil ecosystem and can assist the organic farmer in producing a healthy crop. Micro-organisms can be grouped according to their function: free-living decomposers convert organic matter into nutrients for plants and other micro-organisms, rhizosphere organisms are symbiotically associated with the plant roots and free-living nitrogen fixers.

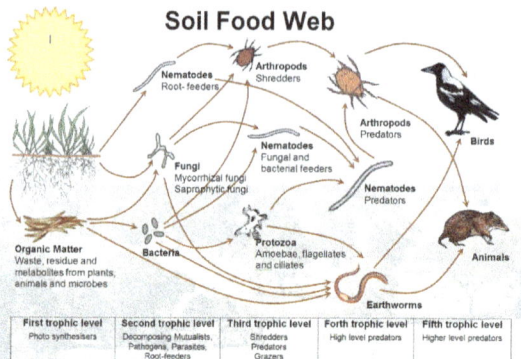

Compare and contrast the structure and properties of sand, clay and loam soils, with reference to a soil texture diagram, including their effect on primary productivity.

Familiarity with the soil texture triangle diagram used for soil type classification based on the percentage of sand, silt, and clay in the soil is required.

Soil structure depends on:

- Soil texture (the amount of sand and clay).

- Dead organic matter.

- Earthworm activity.

For optimum structure, variety of pure sizes are required to allow root prevention, free drainage and water storage. Pore spaces over 0.1 mm allow roots growth, oxygen diffusion and water movement where as pore spaces below 0.5 mm help store water.

Clay:

- Fertile in temperate locations.

- In tropical areas clay is permeable and easily penetrated by roots.

- Nutrient deficient / easily leached in tropics.

The more clay present in soil the higher the force needed to pull a plough.

Different soil types have different levels of primary productivity:

- Sandy soil – low.

- Clay soil – quite low.

- Loam soil – high.

Primary productivity of soil depends on:

- Mineral content.

- Drainage.

- Water-holding capacity.

- Airspaces.

- Biota.

- Potential to hold organic materials.

1. Shrinking limit: state which the soil passes from having a moist to a dry appearance.

2. Plastic limit: occurs when each ped is surrounded by a film of water sufficient to act as a lunricant.

3. Liquid limit: occurs when there is sufficient water to reduce cohesion between the peds.

4. Field capacity: maximum amount of water that a particular soil can hold.

Soil Texture	Nutrient Capacity	Infiltration	Water Holding Capacity	Aeration	Workability
Clay	Good	Poor	Good	Poor	Poor
Slit	Medium	Medium	Medium	Medium	Medium
Sand	Poor	Good	Poor	Good	Good

References

- Soil health_biology, pages, vrosite.nsf, vro, dpi: vic.gov.au, Retrieved January 9, 2019

- Understanding-and-Managing-Soil-Biology: dairyingfortomorrow.com.au, Retrieved January 4, 2019

- Crust: soilcrust.org, Retrieved February 15, 2019

- Dont-bust-biological-soil-crust-preserving-and-restoring-important-desert-resource, rmrs fed.us, Retrieved May 3, 2019

- Ess-topic-51-introduction-to-soil-systems:mrgscience.com, Retrieved January 4, 2019

Permissions

Index

www.ingramcontent.com/pod-product-compliance
Lightning Source LLC
Chambersburg PA
CBHW080407190526
45161CB00003B/160